COTTON & THRIFT

COTTON & THRIFT

FEED SACKS AND THE FABRIC OF AMERICAN HOUSEHOLDS

MARIAN ANN J. MONTGOMERY

FOREWORD BY MERIKAY WALDVOGEL

TEXAS TECH UNIVERSITY PRESS

This book is typeset in Minion Pro. The paper used in this book meets the minimum requirements of ANSI/NISO Z39.48-1992 (R1997). ∞

Photographs of objects in the collections of the Museum of Texas Tech University were taken by Bill Mueller, Julie Lyle, or Jasmine Durrett.

Designed by Kasey McBeath.

Library of Congress Cataloging-in-Publication Data is on file.

ISBN: 978-1-68283-042-0

19 20 21 22 23 24 25 26 27 / 9 8 7 6 5 4 3 2 1

Texas Tech University Press
Box 41037 | Lubbock, Texas 79409-1037 USA
800.832.4042 | ttup@ttu.edu | www.ttupress.org

To the American Quilt Study Group, whose members first showed me the research value and the beauty of feed sacks, and to Anna and Charles Gulden, my maternal grandparents, who raised their family on a Bucks County, Pennsylvania farm.

❧CONTENTS❧

❧FOREWORD ❧

"Make do or do without; do the best with what you've got"—these words became a call to action and a badge of honor for those who lived through the hard times of the Great Depression and the shortages of World War II. Among many stories of "making do," the reuse of feed sacks is remembered with a particular sense of pride.

For homemakers, the lowly cotton sack was a godsend: a free source of fabric to clothe and shelter their families. The National Cotton Council, a trade association, hoped to delay the acceptance of the cheaper paper bag, so textile bag manufacturers devised clever marketing plans to make their sacks attractive. In the early 1930s, added-value premiums--such as doll and puppet patterns—began to be printed on the cloth. Then in the late 1930s manufacturers began printing sack fabric with the same colorful designs as were found on store-bought cloth. To broaden the fabric's appeal to urban and suburban women in the post–WWII era, stylish modern dresses made of feed sack cloth were featured in pattern catalogs with yardage given in number of sacks. As home sewing increased, the use of feed sack cloth lasted well into the 1960s.

As Marian Ann J. Montgomery points out, quilt historians working with quilt documentation projects in the 1980s and 1990s were among the first to begin the research on feed sacks. Feed sack fabrics appeared in many of the quilts brought to quilt days for examination and sharing; when "feed sack prints" are discussed today, most are talking about the sacks printed in the late 1930s that gained widespread popularity. With the heightened interest in 1930s quilts, collecting vintage feed sacks followed, and quilt makers in the twenty-first century bought feed sack fabric to include in their quilts. Japanese quilters in particular paid dearly for imported vintage American feed sacks. Some fabric companies even reproduced the designs in the modern era.

The nineteenth-century introduction of cotton sacks closely followed technological and industrial advances in America. As Montgomery notes, in the 1880s, the cloth sack replaced the wooden barrel as the preferred container for commodity transport and storage because the sack was lighter and took up less space in wagon beds and warehouses. The earliest sacks were made of jute and burlap, but they were eventually replaced by American-grown cotton given pressure from cotton growers in the early twentieth century. Cloth sacks filled with feed and foodstuffs arrived at local stores. Women used the white sacks for curtains, small tablecloths, embroidery projects, and quilt backs. Cotton sack companies eventually noticed how women were adapting the sacks to home-decorating projects and made improvements. They changed the printing ink, for example, when they noticed the difficulty in removing it even after bleaching the sack.

One of the earliest collectors of sacks, garments made from sacks, and publications pertaining to sacks was Anna Lue Cook of Germantown, Tennessee. As passionate as I was about quilts, Cook surpassed my enthusiasm several fold. Her book *Textile Bags Identification and Value Guide: The Feeding and Clothing of America* became my guide to understanding these textiles' complex history, and objects from Cook's collection were included in my book *Soft Covers for Hard Times: Quiltmaking and the Great Depression.* Cook recently passed away, and I am pleased to see her impressive research referenced in this book.

The Pat Nickols Cotton Sack Research Collection, as developed over the last thirty years, presents similar objects important to further research on sack history. The collection includes sacks of all sizes and designs, sacks marked with premiums, garments made from sacks, embroidered household items, and quilts. Also included are twenty-one instructional booklets that outline the types and uses of cotton sacks from 1924–1970. The collection contains over 4,100 feed sack print swatches, and 380 full and partial sacks: many of these collection pieces were photographed for this book, which makes it the most extensive catalog of cotton sack images published to date.

Thirty years ago, a quilt researcher could reach out to quilt owners and families for stories of their quilts—including their memories of feed sack—but voices of that generation are disappearing. What has emerged to replace these first-hand contemporaneous accounts are online collections of digitized newspapers and magazines, from which Montgomery discovered a number of previously unknown aspects of the cotton sack story. For example, an entire chapter in this book is devoted to one of the largest food relief projects ever undertaken: again, the cotton sack played an important, if unexpected, role. Herbert Hoover, prior to becoming President, led this effort to feed an entire nation of people during the First World War. Flour, bagged at local mills in the United States, was shipped by boat to Belgium. Each sack was stamped "From the People of America" alongside the round trademark of the flour mill. These same sacks were later returned to the United States, fascinatingly embroidered with appreciative messages. The Herbert Hoover Home in Iowa and the Hoover Institute at Stanford have several of these embroidered relief sacks in their collections, and one example now also resides in the collection of the Museum of Texas Tech University.

Built upon nearly thirty years of curation and scholarship dating back to the grass-roots efforts of the state quilt projects, *Cotton and Thrift: Feed Sacks and the Fabric of American Households* and the Nickols Collection bring the disparate threads of research together by providing the most comprehensive resource of the history of the cotton sack yet written. The Nickols Collection has already attracted additional donations of sack-related objects—hopefully, this trend will continue and establish further documentation of an important moment in American history.

MERIKAY WALDVOGEL, 2019

☙ DIRECTOR'S NOTE ❧

History is full of wonderful stories, and some of the most interesting are those that did not feature in newspaper headlines. In the exhibition "Cotton and Thrift: Feed Sacks and the Fabric of American Households," and this catalog that supports it, are some truly fascinating stories around one of life's more humble commodities: the cotton feed sack.

Dr. Marian Ann J. Montgomery, Curator of Clothing and Textiles at the Museum of Texas Tech University, has curated the exhibition and written the catalog. She has drawn on the Museum's rich collection of almost 6,000 feed sacks, likely the largest collection of its kind in public hands. How did the Museum come to hold such a collection?

Core to the Museum's remarkable holdings is the Cotton Sack Research Collection assembled by Mrs. Pat L. Nickols. Mrs. Nickols assembled as many different prints of feed sack fabric, as well as types of white cotton sacks and articles made from these sacks, that she could find for her research, which was published in the journal of the American Quilt Study Group, *Uncoverings*. That this significant collection came to the Museum of Texas Tech University in the heart of cotton-growing country is truly fortuitous. Dr. Montgomery was at the right place at the right time and developed a relationship with Mrs. Nickols long before she was ready to part with her research ma-

terials. The Museum is very grateful to Mrs. Nickols for donating a portion of the value of the collection, and we are also grateful to the Quilter's Guild of Dallas, the South Plains Quilt Guild, and several individuals who enabled the Museum to purchase the balance of the collection.

The Nickols Cotton Sack Research Collection comprises over 5600 items ranging from more than 4000 printed feed sack swatches, over 900 whole and partial cotton sacks, 292 quilts, quilt blocks and quilt tops, 120 household articles made from feed sacks, more than 40 garments, 21 instructional booklets from the 1920s to the 1960s promoting the use of cotton sacks, and the very rare embroidered cotton sack used to send flour for hunger relief in Belgium during World War I. While thousands of these sacks were embroidered, painted, or otherwise decorated as thank-yous to Americans for their relief, only a relatively small number survive to this day.

The arrival of the Nickols collection at the Museum spurred others holding feed sack materials to donate them to the Museum, further enriching the collection. These include important donations from Lenna DeMarco, Linda Fisher, Nancy Fehleison, Charlotte Williams, Patricia Hayton, Mari Madison, Dottiemae and Harold Groves, Dan Skarbek, the Ryburn family, and Dr. Robert Bradley. These materials provided Dr. Montgomery with a treasure trove of examples and information through which she has sifted to bring new research to the public in this publication and the associated exhibition.

We are indebted to Mr. Mark Dunn of United Notions/Moda Fabrics for his generous donation and to The CH Foundation for a grant, which together have made it possible to publish this book through Texas Tech University Press and to include many color photographs. We are proud to showcase this little-publicized aspect of American society, one that impacted and reflected the lives of so many people across the nation.

GARY MORGAN, PHD
Executive Director,
Museum of Texas Tech University

⇜A COLLECTION GROWS⇝

Finding feed and flour sacks used as quilt backs in antique quilts piqued my curiosity, which led to my interesting and fruitful study of feed sack quilts. These sacks might have printed mill information as well as the weight and contents of the sack, most commonly held various feed for farm animals or flour needed by farm women as well as city folk for their bread baking.

Quilts, quilt tops, quilt blocks, whole cotton bags--some with printing intact or with paper labels--as well as ads in mostly farm magazines contributed to the growing body of information gathered. When possible these items were purchased to take home, study and start building a database to answer the questions that developed about these sacks. As the number of sack examples grew, so did the number of questions. The answers grew to include many spreadsheets of information, notebooks full of sack swatches, photos of examples of sack fabric in quilts, in garments for the whole family and countless household uses such as dish towels, sheets and pillow cases, curtains and table cloths.

Additional sources of information were booklets from the Home Extension Agent, the Cotton Council, and short farm magazine articles giving homemakers ideas and instructions to utilize sack fabric. Some patterns and instructions were printed directly on the bags; for example, Kent Bags produced a wide variety of specialty bags with printed items to cut out and sew stuffed animals, various toys, and luncheon sets.

This narrative gives you the "why" my large and extensive feed sack collection grew: *curiosity* about what I was seeing.

The story of "how" the collection grew continued in the late 1970s and on into the 1980s. Antique quilts and quilt tops were actively studied as these unfinished items supplied a great deal of information by examining the backs or the wrong sides of the items. Visible was more of the printing, if it had been washed, and often the quiltmaker used the front or the back of the sack. The art of removing the stenciled logos and printing on the bags was quite difficult in the early years: "soak in kerosene" or "boil with lye soap" were two of the suggestions printed on the bag for removing the printing. The presence of printing on the white fabric, either front or back, indicates the homemaker felt it was not a distraction or would hopefully wash out.

There appeared to be mostly two distinct types of cloth: one a larger thread and often coarse weave, the other a smaller thread and finer weave, more like percale or dress goods. There were other examples, but they were much fewer in number, and while interesting they did not warrant in-depth study. The coarse weave sacks held animal feed, with distinction made for the age of the animal: chicken feed was fine grain for baby chickens and changed to more coarse grains as the chicks grew. There was a large listing of all of the various animals that had their feed packaged in cloth bags. The wide variety of seed used for planting new crops also used the coarse bags. Foodstuffs such as flour, sugar, rice, corn meal, oatmeal, and salt were packaged in the finer bags. While today we see only a few cloth bags in most of our grocery stores, red beans, corn meal, and other staples may still be found in specialty grocery stores.

Examples of coarse bags are more easily seen and recognized as feed sack fabric than the finer cloth bags, which resembles percale or purchased dress goods. The easiest way to confirm sack cloth is finding a row of holes, indicating the stitching used to make the bag: the contents pulled at the stitching, creating small holes. Another confirmation of sack fabric is finding an example in one of the collection's binders of sack swatches.

The large number of feed sack (both coarse and fine) examples gathered, possibly over three thousand, approximately 2 ½" x 4" rectangles, were partly sorted into binders, the result of active trading both in the United States and around the world. Members of the Feed Sack Club (no longer active) started exchanging Charm Packs of all different prints. It was such fun to receive a 'squishy' as the puzzled mailmen called them, a fat envelope filled with twenty or more pieces of sack fabric, both feed and flour bags. In trading, I asked for pieces with a row of holes. Since most of those doing the exchanges wanted to use their sack fabric for quilt making and therefore did not want holes, I was welcomed as a trader by those who wanted to get rid of the pieces with holes. There were countless types of images: animals, flowers, childrens' toys, bows, etc., all grouped in binders.

This book's textile history is of great interest to women, especially quilt makers, but we also note the interest expressed by men who remember 'feed sack lore': perhaps their mothers made them a shirt from a feed sack, or perhaps they worked on a farm or in a feed store and handled these heavy sacks. Many remember and can connect with these important aspects of American households.

I am pleased that the exhibit that accompanies this catalog, as well as the ongoing study and research by the Museum of Texas Tech University, will continue to add information about feed sacks.

PAT L. NICKOLS, 2019

ℬℯ ACKNOWLEDGMENTS ℯ℘

One author's name is on this book, but this book was only possible because so many people played a part in providing objects, photographs, research assistance, and funding.

Pat Nickols exhibited great kindness in working with me, and she also demonstrated great patience while I raised the funds needed to get her collection to the Museum of Texas Tech University. Funding to match her donation portion of the value of the collection was generously contributed by The Quilter's Guild of Dallas, Inc., the South Plains Quilt Guild, Linda Fisher, Louise Underwood, Jane West, Pat Grappe, and Carolyn Sowell.

Many staff members of the Museum of Texas Tech University and the Museum Association encouraged and supported this project. Dr. Gary Morgan saw the value of the research collection and encouraged the publication of this book. Dr. Eileen Johnson gave me the opportunity to be Curator of Clothing and Textiles; she encouraged the acquisition of the Nickols collection and the peer-reviewed research that led to this book. Jamie Looney, Jouana Stravlo, and Becky Rogers assisted in securing and distributing the funds for the book's color pictures. Bill Mueller photographed many of the objects and taught Museum Studies graduate students, Julie Lyles and Jasmine Durrett, how to do the rest. The entire exhibits team, who mounted the exhibit that will accompany this book, works tirelessly to help in exhibiting objects from the Clothing and Textiles Division. Graduate students in the Museum Studies Master's Program, who did the hard work of sorting, cataloguing, photographing, and preparing the Nickols and other feed sack collections for exhibition, included Julie Lyles, Jasmine Durrett, Alyssa Stewart, Taylor Ernst, and Cassie Munnell. Other members of the staff contributed in numerous and valuable ways.

As is always the case when funds are sought for a big acquisition, generous donors of similar items come forward. Those that added their treasures to the Museum's holdings or who generously donated the use of their images for this book include: Dr. Robert and Lisa Bradley, Kathleen Hagaman Carson, Leanna DeMarco, Nancy Fehleison and her children, Linda Fisher, Dottiemae and Harold Groves, Patricia Hayton, Virginia Hernandez, Mari Madison, the Ryburn family, Dan Skarbek, and Charlotte Williams.

Members of the American Quilt Study Group provided research assistance, particularly, Merikay Waldvogel and Rose Marie Werner. Annelien van Kempen in the Netherlands directed me to texts written by those who worked in the Belgian relief effort. Margery L. Walker of the National Cotton Council assisted with information on the work of the Council. Jessica Faucher, Corporate Archivist for the General Mills Archives, was helpful in researching the Larro brand and securing images. Being able to access the

Dallas Morning News Archives through the Dallas Public Library website was tremendously helpful.

The publication of so many beautiful color photographs was made possible by contributions from United Notions and Moda Fabrics through its owner Mark Dunn and The CH Foundation.

Joanna Conrad shepherded this manuscript through its journey at the Texas Tech University Press and her cheerful approach kept me calm and the manuscript moving forward.

This work was made easier because of the support of my husband, Barry. He has always encouraged my professional career, even to the point of packing up and moving across his beloved state of Texas so that I could work with the clothing and textiles treasures of the people of Texas that are located at the Museum of Texas Tech University.

MARIAN ANN J. MONTGOMERY, PhD, 2019

COTTON & THRIFT

CHAPTER ONE — INTRODUCTION

Printed cotton sacks are currently fashionable aspects of material culture for research, particularly in the costume and quilt history communities. In the second quarter of the twentieth century, these textiles were a valuable commodity in rural America, serving as a source of free fabric for clothing, quilts, and home décor.

Origin of the Pat L. Nickols Cotton Sack Research Collection

Pat L. Nickols and I first became acquainted as members of the American Quilt Study Group at their annual seminars. In February 2014 Pat contacted me, the newly arrived Curator of Clothing and Textiles at the Museum of Texas Tech University, about the collection of printed cotton sacks that she had amassed for her research into feed sacks. Articles published in *Dress* by the Costume Society of America and *Uncoverings* by the American Quilt Study Group have significantly expanded critical discussion of this topic, and some of Mrs. Nickols's research was published in *Uncoverings* as "The Use of Cotton Sacks in Quiltmaking" (1988). Although she had donated her quilt collection to the Mingei International Museum in San Diego, for the cotton sack collection, Mrs. Nickols sought a purchase/donation. Thus began the adventure that brought over 5,600 pieces of printed cotton sack research materials into the Clothing and Textiles Division of the Museum of Texas Tech University.

This adventure involved raising funds to cover the part of the value of the collection that Mrs. Nickols was not donating. Generous support was provided by a grant from the Quilter's Guild of Dallas, Inc., funds from a donation quilt made by the South Plains Quilt Guild of Lubbock, cash contributions from five individuals, and my donated honorariums from four presentations. These funding efforts resulted in the Nickols collection arriving at the Museum in 2015. Prior to the arrival of the Nickols collection, the Museum held only a few printed cotton sack materials, so the collection increased the holdings dramatically. Additionally, publicity surrounding the fundraising for and acquisition of the Nickols pieces resulted in many more pieces being individually donated to the Museum. Currently the Museum holds almost 6,000 pieces of cotton sack examples, likely making it the largest collection in public hands.

The Pat L. Nickols Cotton Sack Research Collection includes white sacks, printed partial and whole cotton sacks, swatches of printed sacks, instructional booklets, garments, quilts, quilt tops, household objects, and decorated white sacks such as one embroidered in Belgium as a thank-you for food relief. Over 4100 different swatches of printed feed sack fabrics are held in the Nickols collection alone at the Museum of Texas Tech University. Additionally there are over 380 partial sacks with prints and samples of printed cotton sacks from other donors in the Museum's collection. These examples are a significant microcosm of the fabrics that were produced for cotton sacks, and they have great potential for researching quilt, textile, and fashion history as well as providing inspiration for fabric designers.

Given that the Nickols collection was brought to the museum for the purpose of research, it was imperative that a publication accompany the material so that the treasures of the collection could be shared with others. A publication would also help to establish the collection as a venue for future visits from scholars. This publication is therefore intended to serve as the exhibition catalog for the 2019 exhibit of feed sack materials at the Museum of Texas Tech University, to share the highlights of the Pat L. Nickols Printed Cotton Sack Research Collection and examples from other donors, as well as to provide as many visual examples of the prints as possible. The collection is constantly growing: along with the Nickols collection, donations from Lenna DeMarco, Linda Fisher, Nancy Fehleison, Charlotte Williams, Patricia Hayton, Mari Madison, Dottiemae and Harold Groves, Dan Skarbeck, the Ryburn family, and Dr. Robert Bradley add to our understanding of feed sacks. Unless otherwise noted, all objects presented in this book are from the Nickols Collection. Objects in the Nickols Research Collection begin with the catalog number TTU-H2015-053.

Surveying the Research

As is expected from any publication coming from a university, this book has benefitted from the work of earlier scholars, and it is anticipated that this publication will serve future researchers, particularly those not able to visit the Museum's collection in person. Studies by Pat L. Nickols, Ruth Rhoades, and others, published in the American Quilt Study Group's journal, *Uncoverings*, as well as work by Loris Connolly, published in the Costume Society of America's journal, *Dress*, provide basic information about cotton sack manufacturers, their origins in producing white sacks, and how they came to hire designers to create pretty printed fabric to package feeds and staples. Jennifer Lynn Banning, in her 2005 dissertation at Louisiana State University, researched garments made from cotton sacks worn by one woman in rural South Louisiana between 1949 and 1968. Anna Lue Cook's *Identification and Value Guide to Textile Bags (The Feeding and Clothing of America)* documents many different white cotton sacks. Fawn Valentines' West Virginia Quilt documentation, *West Virginia Quilts and Quiltmakers: Echoes From the Hills* provides oral histories that flesh out how the sacks were produced and sold, while Merikay Waldvogel in *Soft Covers for Hard Times: Quiltmaking and the Great Depression* provides an overview of the subject and offers several illustrations of feed sack quilts. Gloria Nixon's *Rag Darlings: Dolls From the Feedsack Era* looks at dolls printed on cotton fabric. Linzee Kull McCray's beautiful *Feed Sacks: The Colourful History of a Frugal Fabric* is a visual treasure. These are valuable resources for further information.

While this volume draws on earlier scholarship, it also adds new research, particularly that related to cotton grown in Texas, and it provides the most extensive visual representation of cotton sacks published to date in telling the story of this American phenomenon. Originating from likely the largest collection of sack research materials held in public hands means that a preponderance of visual examples of printed cotton sacks can be included. Of special importance in identifying feed sack prints are the hundreds of examples of feed sack fabric prints provided in the final chapter of the book, demonstrating the great diversity of printed fabric produced. Thus it is hoped that this publication will serve to gather in one place the wide variety of well-documented information on cotton sacks in America.

A Brief History of Textile Sacks

Textiles both before and sometimes after the industrial revolution have been precious. Fabric for making clothing and for home decoration has historically been difficult to obtain, especially for Americans living on the frontier or in rural areas. Even with the advent of printed catalogs that sold fabric, those living in rural areas were often suspicious of mail-order fabric that sometimes arrived of poor quality, for example, as loosely woven and not colorfast. In the twenty-first century the Internet has had an impact on fabric sales at brick and mortar stores, forcing many stores to also have an online presence; despite this modern-day necessity, consumers still seem to want to touch and see fabric before buying. If given the choice, quilters in particular will want to see the fabric to aid in matching colors.

To rural families in nineteenth century America, the cotton sacks that contained animal feed and staples such as flour were valuable sources of free fabric that could be repurposed. Large households and farms required large quantities of flour, sugar, and other staples of human consumption. Although the term cotton sacks is a generic term, the type of sacks used in this manner changed over time. The earliest cotton sacks were white and printed with logos of the manufacturer; possibly as early as 1927, but certainly[1] after 1937, the sacks were printed with pretty patterns, dolls, and pillowcase motifs with the manufacturer's information on a paper band that was glued to the sack. Crucially, the fabric provided by the cotton sacks was not something a woman had to purchase, either from her own source of funds or with her husband's permission. Rhoades, citing documentation discovered during her Georgia quilt documentation project, states that as early as the 1870s, quilt makers recognized cotton sacks as a source of quilt fabric.[2] White cotton sacks were ubiquitous in rural America from the last half of the nineteenth century, and then sacks printed with pretty designs came on the market in 1937.[3]

While the fabric from the earliest sacks was white and often contained printed logos, its appearance didn't matter for household cleaning, and such sacks were often used for towels and rags. The appearance of printed logos was considerably less desirable when using the fabric for clothing, so women sought ways to remove the inked labels (some methods, as de-

▼ Luther Theodore and Minnie Irene Souder Jeans and family on their farm in Diamond Missouri, circa 1942. Mrs. Jeans and her daughters appear to be wearing garments made of feed sack fabrics. Mrs. Jeans's embroideries on white feed sacks appear in Chapter Three on Decorated Cotton Sacks. Photo courtesy of Dr. Robert and Lisa Bradley.

scribed later in Chapter Five, were more successful than others). Several objects in the Museum's collection retain the ink that women had endeavored to remove, an indication of the difficulty in removing the printed logos and helpful to historians researching the source of the fabric today.

In addition to bags for products for human consumption, beginning in the 1920s animal feed, particularly chicken feed, began being packaged in cotton sacks[4]. As the use of cotton sacks for feed increased, whether the sacks were in actuality intended for human food or other products, the materials themselves became colloquially known as "feed sacks".[5]

What was used before cotton sacks to package these commodities is of interest because it influenced the design of the printed logos stamped onto the sacks. The wooden barrel, properly called the cask, has been in use since the time of the Romans.[6] Historically, all non-living cargo was shipped in a wooden cask, no matter what it was. Today we commonly refer to these items as barrels, but in fact the correct term is "cask" because "barrel" denoted a cask of a specific size. Casks were made of different woods and through a variety of methods depending on whether the casks were intended to store dry or wet goods. Flour casks were made of specific woods bound together in a manner that contained the flour but did not compress it to the point it would explode.[7] Casks were printed with the producer's logo on the round, flat top of the cask.

Until the late 1840s, casks were the primary containers for packing farm and food products.[8] However, casks were cumbersome to move and required much more effort to produce than cotton sacks, which took over packaging staples in the 1850s. While casks fit well in the hold of a ship, the advent of the railroad limited their usefulness since casks had to ride sitting up and took up a great deal more space than cotton sacks, which could lay flat.[9] Additionally, improvements in sewing machine technology allowed for more efficient production of fabric bags in particular, which could now be produced

with strong seams.[10] By the 1890s, cotton sacks had surpassed casks as food packaging.[11]

By 1932, the United States Department of Agriculture estimated that 68 percent of the flour produced in the country was packaged in cotton bags, 21.5 percent in burlap bags, 10.5 percent in paper bags, and less than 0.5 of 1 percent in barrels. In that year, mills generally used burlap bags as large-size containers; cotton bags in all sizes, but most commonly as medium-sized containers; and paper bags as small retail containers. The reusable nature--and hence, the value--of the cotton bag to bakers and housewives was noted.[12]

Most manufacturers didn't discontinue using their round logos, designed for casks, when the switch was made to packaging in cotton bags.[13] This design format persisted until the change to printing an advertising logo on paper bands was introduced: for the paper bands, a logo needed to fit within a rectangular format. Another holdover from the days of casks was that textile bag sizes continued to hold the same amount of their commodity as would a cask. Rhoades records in her article on cotton sacks in Georgia that a 49-pound flour sack held the equivalent of a quarter-barrel, and a 98-pound sack was equivalent to a half-barrel.[14]

Manufacturers responded to the challenges consumers faced in removing the printing from the sacks so that the white fabric that remained could be used for towels, cleaning cloths, diapers, and in some cases undergarments. The inks that were used for the logo changed over time, with manufacturers making the logos easier and easier to remove until, by the end of the 1930s, the practice became to place logos onto printed paper bands glued to the sack.

Cotton was in competition with paper sacks as early as the middle of the nineteenth century. The shortage of cotton during the Civil War forced more paper packaging for sacks. After the war, both cotton and paper sacks were used. In 1889 C. B. Stilwell patented the process that created flat bottom paper bags, which continue to be used today for sugar and flour.[15] By the early twentieth century, paper

▲ This printed cotton sack, printed circa 1940, retains the central round logo as the focus of the label and was likely an early label when the graphic designers were still exploring the new medium of rectangular paper, TTU-H2015-053-003-266.

▲ The printed label on this feed sack, circa 1920, uses the round logo that would have also worked on the top of a cask or barrel. Partial gift of Pat L. Nickols and funds from The Quilter's Guild of Dallas, Inc., The South Plains Quilter's Guild and individuals, TTU-H2015-053-003-354.

▶ This printed cotton sack, printed circa 1940, retains its original rectangular-shaped paper label, which was developed to make it easier for a homemaker to take the label off the bag and have all the fabric to use, TTU-H2015-053-003-907.

was gaining market share because it was considered more sanitary and cost efficient. However, Rhoades records at least three prominent textile bag manufacturers at the turn of the century: Bemis Company, which originated in St. Louis, Missouri, in 1858; Chase Bag Company, founded in Boston, Massachusetts, in 1847; and the Fulton Bag and Cotton Mills of Atlanta, Georgia, founded in 1881.[16] Cook's research in *Davison's Textile Blue Book* documents 24 textile mills manufacturing bag goods in 1932, 31 in 1942, and 33 in 1952.[17] In addition to the manufacturers named by Rhodes, Cook lists Alabama Mills Company, of Birmingham, Alabama; Royal River Mills of Yarmouth, Maine; Laurel Mills of Laurel, Mississippi; Flint River Cotton Mills of Albany, Georgia; Illinois State Penitentiary of Joliet, Illinois; Cannon Mills Company of Kannapolis, North Carolina; Lone Star Cotton Mills of El Paso, Texas; and Harmony Grove Mills of Commerce, Georgia.[18] Their manufacturing was aided by a surplus of cotton in the 1920s, the introduction of the mechanical cotton picker in 1936, and the promotional efforts of the cotton industry.

In the earliest days of cotton bags that were used for food or grain, the bag was unfinished greige-goods of osnaburg or unbleached muslin.[19] Burlap bags were also used, but they were usually reserved for dairy, horse, and mule feed.[20] The burlap bags were coarse and not reusable for as many purposes as were the white bags. According to Hoye who wrote *Staple Cotton Fabrics* in 1942, fabrics commonly used for cotton sacks were print cloth sheeting (combed cotton in plain weave), osnaburg, and gingham. Of these, Hoye cites plain cloth in the greige (gray) state, no finish or print added, as being used in very large quantities by the bag trade for grain and feed bags.[21]

In October of 1924, Asa T. Bales took out a patent for the George P. Plant Milling Company of St. Louis, Missouri, for making bags out of fabric suitable for clothing. This was the source of the fabric for the Gingham Girl Flour, which began packaging their flour in red and white gingham fabric that year.[22] Mother's Gingham Flour also packaged their

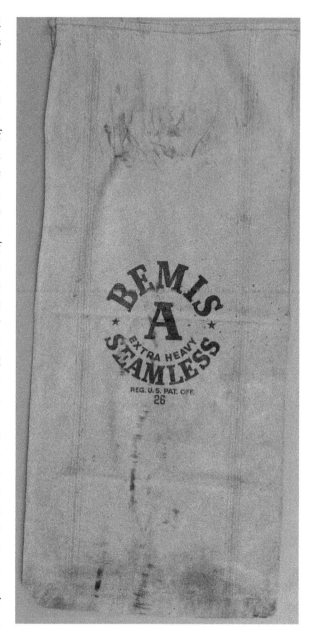

▲ Bemis-labeled bag with round logo, circa 1910, TTU-H2015-053-003-128.

flour in a pink and white gingham pattern. For some reason they did not expand beyond this one small gingham design. Cotton sack manufacturing companies like the George P. Plant Milling Company prospered even through the Depression years because, firstly, families needed the products contained in the sacks,[23] and secondly, when offered a choice, the family often chose the product in a cotton sack over a paper sack because the sack could be used as a source of free fabric.

Depression-Era Cotton Production

Companies such as the Imperial Sugar Company of Sugarland, Texas, and the Robert Nicholson Seed Company of Dallas, Texas, switched from jute and burlap bags respectively to cotton in order to help ease the cotton slump that hit farmers throughout the South in the early 1920s. Estimates in 1926 were that if cotton was used to replace jute in textile bags, about 4,000,000 more bales of cotton per year would find a purpose, significantly easing the cotton farmer's difficulties.[24] Burlap was cheaper to use than cotton, but it was found that cotton bags could be reused more times than burlap with the higher cost offset by the number of uses. When the price of cotton was down, the bags were even more reasonable.[25] In 1926 the Farish Company of New York developed a research department for the purpose of seeking out new uses for cotton to increase consumption. An effort was made to produce a cotton burlap that would replace jute burlaps. Jute was imported—often from India—to make burlap. If cotton could replace jute, it was estimated that it would use 1,300,000 bales of cotton, keeping 2,600,00 cotton spindles active year round. Cotton bags were thought to look cleaner than jute bags, could be better branded, and would have many uses after they were empty that jute did not have.[26] In 1927 the Bureau of Agriculture Economics of the Department of Agriculture estimated that if cotton bags were used in place of bags made out of other materials for the wholesale grocery trade, that 750,000 more bales of cotton would be used.[27] In 1930 the *Dallas Morning News* reported that there was an increasing trend toward packaging of foodstuffs in cotton bags, with the sales of these bags being nearly five times as large in 1929 as in 1928. They also reported the advantages of cot-

▲ Gingham Girl Flour bag in gingham, suitable for clothing, 1926, TTU-H2015-053-003-575.

▲ Example of gingham fabric in Nickols Collection, pink and white, circa 1940, TTU-H2015-053-003-899.

ton bags in making more attractive containers and affording an opportunity for distinctive marking of individual brand names. [28]

The battle to increase the use of cotton was particularly important to those living in cotton-growing areas such as Texas. Cotton was inextricably linked with the business prosperity of Texas, so the cotton surplus of the mid-to-late 1920s caused cotton producers and others to look for additional ways to use cotton.[29] The continuing surplus and the effects of the Depression resulted in marketing campaigns encouraging people to use more cotton, such as buying cotton clothes and demanding that groceries be packaged in cotton bags in an effort to provide income for those who had been impoverished by economic factors the South had been unable to control.[30] In 1931 the Wyatt Food Stores, a Dallas organization, announced that they were purchasing a railroad car's worth of Skinner's macaroni and spaghetti to sell in Dallas, stating that the main reason was Skinner's new policy of using cotton cloth bags to package the product. This was called "the Dixie Package." Skinner's sack was printed in red, white, and blue and had "received the approval of intelligent women buyers everywhere who already know the quality of Skinner's products . . .[and] loyal Dallas women are co-operating in every way possible to increase the various uses for the consumption of cotton, thus aiding in a very material way the return of better times."[31] The Wyatt Food Stores expanded their support of the Texas economy in 1935 by conceiving of the idea of putting sugar in cotton bags. Additional benefits cited by the company were that the bags were more sanitary and looked better.[32]

The Bureau of Agricultural Economics, a division of the U.S. Department of Agriculture, experimented with cotton for sugar sacks in 1936. While jute bags were cheaper, as mentioned previously, the fiber came from India and didn't help the American economy. Cotton sugar bags could be used an estimated three times as opposed to jute bags, which were only used once. The Hawaiian Sugar Association tested the use

of cotton bags to ship raw sugar from Hawaii to California: it was estimated that this new use provided an outlet for 50,000 more bales of cotton a year.[33] This was a small part of the cotton surplus that existed at the time, but it was thought that every new opportunity to use cotton was important.[34] By 1933, the Cotton Textile Institute was reporting that cotton bags were used to package 500 products, requiring hundreds of millions of cotton bags a year. Among the products shipped in cotton sacks were potatoes, oranges, and other fruit and produce.[35] Sugar alone accounted for a significant use of cotton fabric, with one company having used 40,000,000 yards of sheeting in making 99,000,000 cotton sugar bags in 1932, and 9,000,000 yards of sheeting for cotton liners. At the same time, 12,000,000 potato sacks were made of cotton.[36] In a 1939 article published in the *Dallas Morning News*, Mark Anthony, president of the Dallas Shippers Association and one of the Southwest's leading cotton men, touted the use of these textiles, saying, "The use of more cotton bags in place of paper, burlap and jute would go far in stimulating domestic consumption of Texas' greatest cash crop, putting more money in farmers' pockets, which means putting more money in everybody's pockets." [37] Anthony went on to say that every time a railroad car-load of sugar was packed in cotton bags, it used up 16,000 five-pound cotton pockets and 800 cotton containers, equaling the production of four and two-thirds acres of cotton land; by that logic, the car-load of cotton-bagged sugar gave a day's work to 43 cotton farmers, 32 cotton mill employees, and 8 cotton bag employees.[38]

Other Texas citizens contributed to this concerted effort to aid the Texas economy, such as the Texas Sheep and Goat Raisers' Association, which received cotton bags for their wool as a gift of the Surplus Commodity Corporation of the United States Department of Agriculture, which was trying to find a new outlet for cotton. The bags they had been using were made of imported jute.[39] In 1938 Texas Commissioner of Agriculture telegraphed Federal authorities to protest their proposal to use paper instead of cotton bags for packing 1,000,000 barrels of flour that the Federal Surplus Commodities Corporation would purchase. The Commissioner estimated that the use of cotton bags would mean sales of about 2,000 bales of cotton.[40] Officials in other cotton producing states, such as Alabama, also worked to have more cotton bags used. In 1927 the Alabama State Department of Agriculture tested the practicality of shipping fertilizer in osnaburg sacks, made from cotton. The experiment showed that the cotton bags could effectively carry fertilizer, providing another use for the fiber.[41] In the late 1940s Fulton Bag and Cotton Mills developed a coating that was put on cotton bags for seed grain, livestock feeds, and other grain not for human consumption that repelled rats and therefore provided a safer environment for the shipment of grain. The coating cost approximately a penny a bag, but the cost was easily off set by the industry-estimated $500,000,000 per year that was lost to rat damage. The chemical was soluble, so if a woman wanted the bag for its fabric, the poison could be removed with a washing.

The debate about whether to package onions in cotton or paper bags ran for years, with Texas cotton producers and the National Cotton Council encouraging farmers growing onions to use cotton bags because doing so would provide thousands of hours of work to laborers in Southern cotton mills.[42] By 1941 woven cotton mesh bags were increasingly being used to ship the Texas onion crop because improved manufacturing processes enabled the bags to be dyed in more brilliant colors than paper bags. These attractive cotton packages resulted in sales of onions in Eastern markets at premium prices.[43]

Promotion and Marketing Efforts

Another aspect of promoting the use of cotton in sacks were those that printed toys and dolls on the fabrics, which could be cut out and assembled. Commercially made cut-and-sew dolls and toys were in-troduced during the 1880s.[44] Manufacturers seized on their popularity and began to use them in advertising, offering them as premiums.[45] According to Nixon, early advertising dolls were printed on muslin or linen sheets and mailed to the consumer. Advertising dolls were created by several different companies. Prior to 1937, these dolls were acquired by sending in coupons. Sometimes the company name was printed on the doll. Later the dolls were printed on the flour and feed sacks.[46] Pertinent to this book is the fact that sometimes the early two-dimensional stuffed cotton dolls are confused with those that were printed on the bags after 1937 because they are of similar construction. To clear up the confusion, a brief look at the early dolls in the Museum's collection follows.

The Aunt Jemima doll is believed to be the first cut-and-sew advertising doll offered by a flour miller.[47] Advertisements for the doll ran in newspapers in 1905, and the following year a doll coupon was printed on boxes of pancake flour.[48] By 1909, a full line of dolls from Aunt Jemima's family were offered for a small payment accompanied by box tops.[49] Although of the cut-and-sew construction which would be used after 1937 on cotton sacks, the Uncle Mose doll currently in the Museum's collection was not printed on a cotton sack. The print ads that offered these dolls with a small payment accompanied by box tops from Aunt Jemima products document that these items were not actually from cotton sacks packaging the Aunt Jemima product. Images in Nixon's *Rag Darlings: Dolls From the Feedsack Era* date the Museum's piece from 1923.[50]

In 1928 Kellogg developed their Lithograph Fairyland Series of dolls that included Red Riding Hood and Little Bo Peep. These dolls were offered as a premium/coupon on Kellogg's Corn Flakes and Pep cereal boxes. The Little Red Riding Hood doll carried a basket of Kellogg's products including a box of cornflakes and a box of pep. The Museum's collection includes a Little Bo Peep cloth doll from this series.

▲ 1916 ad for Aunt Jemima's Pancake Flour with special offer for the Rag Doll Family, Private collection.

▲ 1925 print ad for Aunt Jemima Pancake Flour included a coupon to mail-in for the Jolly rag dolls, which included Uncle Mose, an example of which is in the Museum's collection, Private collection.

▲ December 1924 ad for Aunt Jemima Pancake Flour with special offer for the Aunt Jemima family rag doll family, including Uncle Mose. Private collection.

In December 1934 Eagle Milling of Edmond, Oklahoma, printed the first in its series of dolls on its sacks of Rag Darling Flour.[51] The Museum's example shown on p. 14, is the first in the series, depicting a doll in a white dress trimmed in red. On the reverse of the bag, the back of the doll and fabric for the soles of her shoes are printed. Additionally, for their promotions around the California Pacific International Exposition in 1935 and 1936, the Sea Island Sugar Company began printing dolls from a variety of countries on their bags. The availability of the dolls was promoted in newspaper ads. These dolls were printed in seven sets of five over a two-year period.

Following the success of Rodkey's Rag Darling and the Sea Island Sugar dolls, other mills began pro-

▲ Aunt Jemima's husband, Uncle Mose doll, circa 1923. Gift of Mrs. Tom Masterson, TTU-H1964-083-002d.

▲ 1928 Kellogg Little Bo Peep doll. Gift of H. A. Anderson, TTU-H1966-004-004.

ducing their own dolls and toys. Examples from both the Percy Kent Bag Company and the Fulton Bag and Cotton Mills are represented in the Museum's collection, shown on p. 15.

Cotton Sacks Gain Popularity for Household Use

In the 1930s the cotton sack industry developed a new technique for retaining consumer demand for cotton sack packaging. In 1933 Secretary of Agriculture Henry A. Wallace promoted a variety of dresses made out of sugar bags and chicken feed bags during his trip through the Cotton Belt.[52] These would have been made from white sacks, which were white but had been home dyed, or were from white fabric trimmed with purchased printed fabric. The cost of the dresses ranged downward to .03 cents, and pho-

tographs of society women wearing the dresses were run in the newspapers.[53] Rhoades mentions that by the 1870s some cotton bag manufacturers offered bags in three or four colors (however, these being solid colors makes it extremely difficult to find remaining evidence of this statement).[54] In 1936 Staley Milling Company of Kansas City offered "Tint-sax," which were pastel-colored bags in a fine dress goods fabric that were offered in eleven different shades and manufactured by the Percy Kent Bag Company.[55] These were solid-colored bags and not prints.

However, when printed cotton sacks are discussed today, most remember the beautiful printed sacks developed after 1937 that became a huge phenomenon. It is believed that Richard K. Peek, Vice President of the Percy Kent Bag Company, came up with the idea of printing the cotton bags so that they could be used for home decoration and clothing in

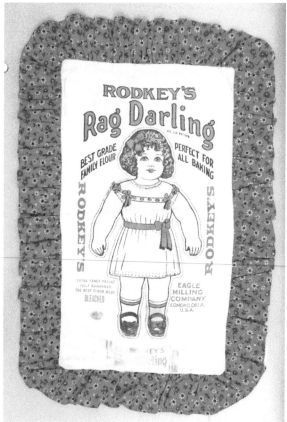

▲ An example of the first Rodkey's Flour Rag Darling printed on a sack from Eagle Milling, Edmond, Oklahoma, 1934, TTU-H2015-053-004-036.

▲ Tanya the Russian Girl from the set published June 12, 1935.

▲ Pedro the Mexican Boy from the set published September 15, 1935.

◄ Little girl in red polka dot dress doll made from a feed sack, circa 1940. Gift of Mrs. Lou Pruitt, TTU-H1968-034-012.

▲ Directions on Percy Kent Bag Company bag for making a Humpty Dumpty doll, TTU-H2015-053-003-315, detail.

◄ Percy Kent Bag Company bag for Central Flour Company. The sack back has pattern for making a Humpty Dumpty doll, circa 1940, TTU-H2015-053-003-315.

▲ Fulton Bag and Cotton Mills bag for Southern Flour Mills self-rising flour with a pattern on the back for making a sailor doll, circa 1940, TTU-H2015-053-003-309.

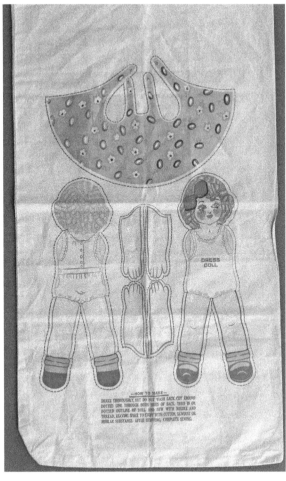

► Fulton Bag and Cotton Mills bag for Southern Flour Mills self-rising flour with a pattern on the back for making a doll to stuff and a dress for it to wear, circa 1950, TTU-H2015-053-003-305.

▲ Tint-sax bag (front and reverse) created by the Percy Kent Bag Company. The Tint-sax logo appears in red at the top of the bag. The back of the bag is printed so that it can be embroidered and made into an apron, 1938, TTU-H2015-053-003-310.

1937.[56] This idea led to the multitude of floral, stripe, plaid, geometric, abstract, and conversation prints that were developed and used on cotton sacks. The fabric was very desirable, with the headline on page one of the July 1944 *Dallas Morning News* proclaiming, "Why Not This Paisley Pattern? Biddy Won't Know Difference." The article went on to describe the cotton bags' popularity in rural households:

Thousands of farm wives in Texas and all over the nation . . . buy their chicken feed by the colors and designs on the bag rather than by the contents. The bags . . . are made of good quality print goods, from which the wives make colorful house dresses for themselves and their daughters. . . .The fabrics appeared more like dress goods rather than containers for such things as chicken feed. Prints in checks, flowered patterns and nursery designs as attractive as one could find on any dry goods counter lure the farm wives, and they make careful checks of patterns and designs available in their feed stores before they buy.[57]

A meeting of the Cotton Research Congress in Dallas in 1949 was the opportunity to promote "Flour Sack Fashions." The promotional article described a fashion show featuring various garments created from the cotton sack material:

> In fashion, it's not what you use but how you use it. Examples are these dresses to be seen in the all-cotton-bag fashion Thursday night. . . . These two young Dallas matrons, accustomed to modeling silks and satins, lose none of their runway aplomb in the afternoon dresses from printed cotton flour sacks. The show . . . will also include a 1-sack swimsuit made from a single 100-pound bag, a fringed sun dress from fertilizer bags, evening dresses from printed cotton flour sacks and square –dance ensembles showing four cotton feed sack dresses.[58]

In 1952 the *Dallas Morning News*, under the headline "Bag Clothing Again," showcased cotton sack fashions. The article noted these fashions' continued popularity, observing, "Just to show what can be done with bags produced for the feed flour and fertilizer industries, the National Cotton Council exhibited women's clothing made from them. For several years the bags have carried designs making it possible for farmers' wives to convert the cloth."[59] The focus here was still on selling the bags to farmer's wives, women living in rural areas. The popularity of this readily accessible source of domestic fabric was continuing to endure.

▲ Example of printed cotton sacks that became available after 1937 on a flour sack from the Kimbell-Diamond Milling Company of Fort Worth, Texas. Gift of Mari Madison, TTU-H2018-008-003.

CHAPTER TWO — WHITE COTTON SACKS; PRINTED DETAILS GIVE CLUES TO PLACE IN HISTORY

The earliest cotton sacks were white and were printed with the logo of the firm that produced the product in the sack. Pretty prints were added to sacks in 1937. The focus of this chapter is the white sacks, along with how legislation of the time provides clues today to the date of the sack.

Rural housewives valued the white fabric in cotton sacks as free fabric for towels, rags, baby clothing, and undergarments.[1] The 100-pound bags generally measured 42 inches wide and 36 inches long, which was more than a square yard of white fabric. The artwork on the bags often featured eye-catching logos with a prominent circle in the middle of many of the logos, a holdover from the round logos used on the barrel lids. The companies did not redesign their logos when they switched from barrels to sacks: they kept the same logo shape. There are almost 200 examples of white bags in the Nickols collection. Not all of their images could be included in this book, but some of the most interesting follow.

The white bags are interesting historical documents. Cook states that in the late 1920s the seed companies started to buy good toweling bags in an effort to obtain a package that would be consumed, utilized, and attractive to the housewife.[2] An example of this fabric is seen in the seed bag, shown on p. 20, TTU-H2015-053-003-097. Additionally, the white bags have historical value because the labels that were printed on the white feed sacks required specific language as new laws were enacted. Text printed on one bag states, "Notice—we have taken great care in growing or selecting and thoroughly cleaning this seed and believe that it is true to name and complies strictly to all requirements of law as to purity and vitality, but we want it distinctly understood that we do not give any warranty expressed or implied. If the purchaser does not accept the seeds on these terms and conditions, they are to be returned at once." This bag was likely printed in response to the enactment of the Federal Seed Act, which went into effect on February 4, 1940, and required accurate labeling and purity standards for seeds in commerce and prohibited the importation and movement of adulterated or misbranded seeds.

The text on Larro bags, produced by General Mills, documents when the products were made.

This bag contains the required notice for the Federal Seed Act, but the colored woven strip in this bag made it particularly desirable for use as a towel, once the logo was removed, circa 1940, TTU-H2015-053-003-097.

▶ Printed for The Attala Company in Kosciuski, Mississippi, this colorful bag required two passes of color to be printed. Kosciuski is named for the U.S. and Polish-Lithuanian general Tadeus Kosciuszko who participated in the American Revolution. The town is also famous as the birthplace of Oprah Winfrey and James Meredith, circa 1930, TTU-H2015-053-003-052.

◄ This bag uses the outline of the state of Alabama as part of its marketing for Bama Laying Mash, circa 1930, TTU-H2015-053-003-032.

► This white cotton sugar sack,1933-35, bears the logo of the National Recovery Administration (NRA). The NRA was set up as part of the New Deal, in 1933, and its goal was to eliminate cutthroat competition by bringing industry, labor, and government together to create codes of fair practices and set prices. In 1935 the U. S. Supreme Court declared the NRA unconstitutional because it infringed upon the separation of powers, particularly Congress's role to pass legislation. However, after the NRA was declared unconstitutional, Congress passed legislation that effectively accomplished what the NRA was created to do. Gift of Dottiemae and Harold Groves, TTU-H2018-018-001.

▲ Larro ad from March 1943 includes an image of bag with "Farm-tested" on it. Notice the round logo and also the information on how to pronounce "Larro" to rhyme with "Arrow." Courtesy General Mills Archives.

During World War II ingredient shortages and government restrictions upon formulas made modifications of the feed necessary.[3] Larro had prided itself on its secondary trademark, "Farm Tested," which appeared on its bags promoting the quality of the product that had been refined on their test farm. By the end of 1943, "Farm Tested" was taken off the bag and instead bags were printed with the sentence, "This feed represents our sincere effort to give you the best value under present conditions." It is rare to be able to date a white sack, but the statement, "This feed represents our sincere effort to give you the best value under present conditions," dates any bag bear-

ing it to after December 1943[4] and before December 1946.[5]

One additional clue to the dating of a white sack is the standardization of product amounts that came at about the time of World War II. Prior to this time, sacks had held the amounts that were standard for barrels. In some cases this meant weights of commodities, such as 24 or 48 pounds. In 1943, during World War II, these weights were standardized so that commodities came in 5-, 10-, 20-, 25-, and 50-pound bags.[6]

▲ This Larro Grains bag was produced during WWII when the feed was altered as part of the rationing plan. Notice the statement, "This feed represents our sincere effort to give you the best value under present conditions…," and that the phrase "Farm-tested" does not appear on the bag. This dates the bag between December 1943 and December 1946. TTU-H2015-053-003-095.

◄ Ad placed in trade magazines for Larro feeds, likely dated after December 1946 because the sacks have prints and "Farm-tested" on them. Notice the continued use of the round Larro logo. Courtesy General Mills Archives.

Prior to the production of the beautiful prints on cotton sacks, the companies also printed the bags with blocks that could be made into quilts or embroidered (see p. 41–44).

Although they might seem difficult to document with the date they were produced, white cotton sacks sometimes bear clues to when they were made based on legislation and relief efforts of the time. These cotton sacks provided good, useable white fabric that the homemaker could reuse for household items.

◄ Photo of boy feeding calf from the Larro Calf Builder feed sack. The sack bears the "Farm-tested" trademark, dating it prior to December 1943 or after December 1946. Courtesy of General Mills Archives.

◄ This bag has printed details that give a clue to its place in history. It could be returned for credit and references rations: "Quality Home Rations Hemp Seed." It also references relief efforts: "One of the United Nations, from the United States of America." This places the bag after WW II, when the UN first provided humanitarian aid to the devastated continent of Europe, and likely before 1961, when the World Food Programme was established as a division of the United Nations. The upper left-hand box provides directions for how to remove the ink from the bag: "Soak in cold water with soap powder overnight. Scrub with hot water in morning." TTU-H2015-053-003-117.

► Some bags bear images of animals, such as this one for rabbit pellets with the rabbit motif around the border and a rabbit enjoying a grassy county meadow, circa 1930, TTU-H2015-053-003-101.

◀ This cotton sack was printed with a beautiful fishing scene for the Lavonia Roller Mill of Lavonia, Georgia, circa 1930, TTU-H2015-053-003-116.

▶ This bag includes hens, still in a circle that harkens back to the original round barrel logos, on growing mash from Taylor Grain Company of Van Alstyne, Texas, circa 1930. Containing only 8 1/3 pounds of product, the bag measures 9 ½ inches wide x 17 ½ inches long. TTU-H2015-053-003-256.

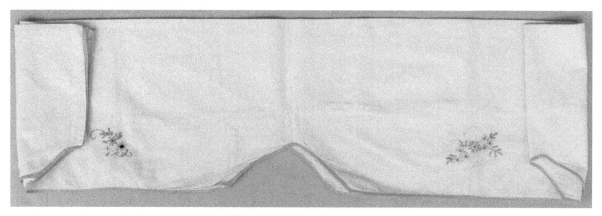

▲ A chin guard was made to protect the upper edge of a quilt from wear. This is one was made from a cotton sack, trimmed with purchased bias binding and embroidered, circa 1925, TTU-H2015-053-004-028.

▲ This valence was made from a white cotton sack that was appliqued with printed cotton feed sacks and embroidered, c 1940, TTU-H2015-053-004-022.

◄ Clothespin bag with a Sunbonnet Sue figure, made from a white cotton sack and trimmed and decorated with printed cotton feed sack fabric circa 1945, TTU-H2015-053-004-035.

CHAPTER THREE ⁓ DECORATION
AND USE OF PLAIN WHITE SACKS

Cotton sacks were recycled into dishtowels and cleaning rags beginning at the end of the nineteenth century. Early in the twentieth century, it became popular to embroider the cotton sacks. The Nickols collection contains several examples of embroidered white cotton feed sack fabric. The items in the Nickols collection all come without information on who made the objects. Additionally, the Museum has embroidered cotton sacks from other donors for whom the maker is known.

Clothespin bags were very useful when homemakers hung their laundry on a line to dry. Several examples of this utilitarian object exist in the Nickols Cotton Sack Research Collection (see p. 28).

In rural areas, household linens were often made from cotton sacks and embroidered with patterns. While there is no published source or reference archive documenting which embroidery patterns were published by what company, there are a few areas that have been researched that aid in understanding when the pieces were made. Prior to the introduction of iron-on transfers for embroidery patterns, there were perforated patterns that were used with pow-dered ink, which was sprinkled through the holes of the pattern to mark the design on the fabric.[1] Gunn records that the heat-transfer process had been developed in England in the late nineteenth century by William Briggs and Co., and in 1902 the Kaumagraph Company had established a production facility for the process in Newark, New Jersey. In 1912 the McCall Company engaged the Kaumagraph Company to print heat-transfer designs for McCall's needlework patterns.[2] According to embroidery scholar Rose Marie Werner, Aunt Martha offered iron-on transfers in the 1930s and Vogart in the 1940s.[3]

The towels on p. 30 that are decorated with rickrack likely date to sometime after the Depression, when there was time to do more involved embroidery and more money available to add rickrack.

According to Werner, there were many embroidery patterns for dish towel sets that took people or animals through the process of courtship, marriage, and child rearing with cute, rhyming captions. The "Cute Catch" and "Old Batch" towels pictured on p. 31 are likely part of such a set.[4]

▶ Clothespin bag designed with Overall Sam figure, made of a white cotton sack appliqued with printed cotton feed sacks and embroidered, circa1945, TTU-H2015-053-004-023.

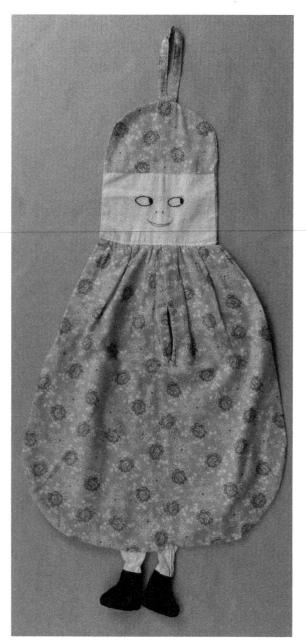

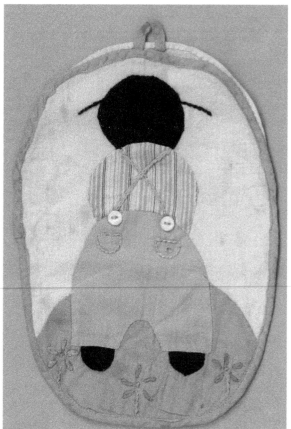

▲ Clothespin bag made of printed cotton feed sacks and embroidered with a face. The piece has legs and shoes attached at the bottom, circa1945, TTU-H2015-053-004-103.

▲ Luncheon cloths, which were small tablecloths designed to fit over a card-table-size table, could be made from a feed sack. This example is made of a feed sack embroidered with yellow flowers, circa 1925, TTU-H2015-053-004-109.

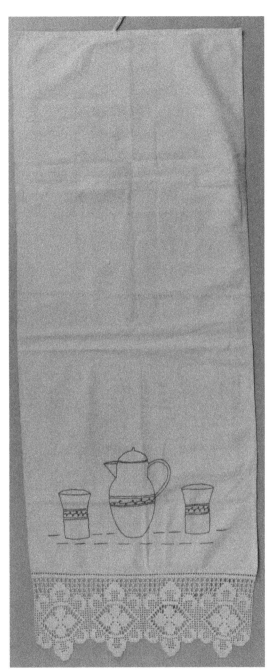

◄ Among the pieces in the Nickols collection is this cotton sack made into a tea towel, embroidered in red with an image of a pitcher and glasses and embellished at the bottom with handmade lace, circa 1920, TTU-H2015-053-004-065.

▲ This white sack was embroidered in cross-stitch and edged with a blanket stitch to make a towel, circa 1925, TTU-H2015-053-004-034.

▲ Hand towel with embroidered rooster, decorated with rick-rack, circa 1945 TTU-H2015-053-004-004.

▲ Hand towel with embroidered hen, decorated with rick-rack, circa 1945, TTU-H2015-053-004-005.

▲ Hand towel with embroidered rooster, decorated with rick-rack, circa 1945, TTU-H2015-053-004-006.

◄ Hand towel with embroidered angel, circa 1945, TTU-H2015-053-004-007.

◄ Hand towel with embroidered rooster among flowers, circa 1930, TTU-H2015-053-004-015.

◄ Hand towel embroidered with dancing rooster and words, "Old Batch," circa 1930, TTU-H2015-053-004-014.

Two sets of embroidered cotton sack towels came to the Museum of Texas Tech University with documentation regarding the maker. Minnie Jeans (Mrs. Luther Theodore, b. Jan. 19, 1916; d. April 25, 1989) of Diamond, Missouri, embroidered some of the nicest pieces. Mrs. Jeans was a farmer's wife who kept busy raising five children and performing household chores. Later in her life, she was a cook for the Diamond, Missouri School System. In addition to embroidery, she enjoyed canning and baking.

In addition to the embroidered cotton sack towels with rabbits, Minnie Jeans also embroidered the popular day-of-the-week motifs onto cotton sack towels (see pp. 34–35). The days of the week towels represented the type of work the housewife was to do each day of the week. While some of the sets switch around the shopping, baking, and cleaning days at the end of the week, all of the sets designate Sunday for church attendance, Monday for washing, Tuesday for ironing, and Wednesday for mending or sewing. The pieces embroidered by Minnie Jeans are from patterns currently available on the Internet at www.tipnut.com, whose author has a set in her collection she dates from the 1940s.[5] Embroidery scholar Werner agrees with the 1940s date, citing the appearance of an electric iron, a dress form, a portable sewing machine, and a carpet sweeper, items which would date from that time.[6]

▲ Minnie Jeans with her family, circa 1950-1952. Photo courtesy Dr. Robert and Lisa Bradley.

◄ Embroidered design by Minnie Jeans of rabbit looking into cookie jar on cotton sack towel, circa 1940. Gift of Dr. Robert Bradley, TTU-H2016-022-012.

◄ Cotton sack towel embroidered by Minnie Jeans with a rabbit holding a carrot, circa 1940. Gift of Dr. Robert Bradley, TTU-H2016-022-013.

◄ Embroidered cotton sack day-of-the-week towel (detail) made by Minnie Jeans representing Sunday, the day to attend church, circa 1940. Gift of Dr. Robert Bradley, TTU-H2016-022-005.

▲ Embroidered cotton sack towel (detail) by Minnie Jeans representing Monday, washday, circa 1940. Gift of Dr. Robert Bradley, TTU-H2016-022-006.

▲ Embroidered cotton sack towel (detail) by Minnie Jeans representing Tuesday, the day to iron everything that had been washed on Monday so that it would not mildew, circa 1940. Gift of Dr. Robert Bradley, TTU-H2016-022-007.

▲ Embroidered cotton sack towel (detail) by Minnie Jeans representing Wednesday, the day to sew, circa 1940. Gift of Dr. Robert Bradley, TTU-H2016-022-008.

▲ Embroidered cotton sack towel (detail) by Minnie Jeans representing Thursday, the day to go grocery shopping, circa 1940. Gift of Dr. Robert Bradley, TTU-H2016-022-009.

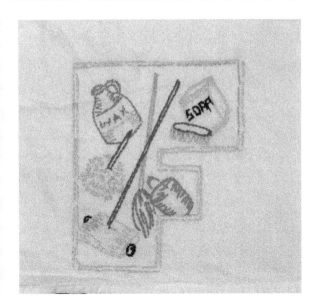

▲ Embroidered cotton sack towel (detail) by Minnie Jeans representing Friday, the day to clean the house, circa 1940. Gift of Dr. Robert Bradley, TTU-H2016-022-010.

The chores assigned to the different days of the week sometimes changed, but, again, Sunday for worship, Monday for washing clothes, Tuesday for ironing clothes, and Wednesday for mending or sewing remained the same.

The Nickols collection has two different day-of-the-week sets, one that features chicks doing the chores for different days of the week (see pp. 36–37). Embroidery scholar Werner, identified the source of the pattern from Modern Handcraft, which was one of the names of the Colonial Pattern Company, part of the Aunt Martha brand. The chick tea towel motifs are #9171, and the title of the undated catalog is "Aunt Ellen's Favorite Designs for Gifts, Bazaars and the Home."[7]

The second group of towels in the Nickols collection (see p. 38) includes only a few days of the week and is based on the Aunt Jemima / Mammy figure that was popular in many formats, from saltshakers to napkin holders. Today we might not likely make objects with this image because of the associated negative connotations.

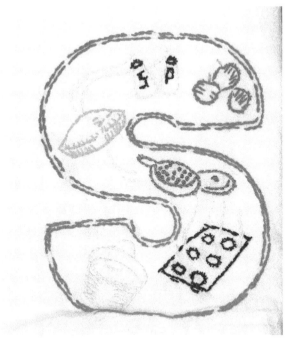

▲ Embroidered cotton sack towel (detail) by Minnie Jeans representing Saturday, the day to bake, circa 1940. Gift of Dr. Robert Bradley, TTU-H2016-022-011.

In this set, Sunday is once again set aside for attending worship, Tuesday for ironing, Wednesday for sewing or mending, but Thursday is for visiting, and Saturday for cleaning.

The Museum of Texas Tech University holds another set of day-of-the-week towels, made by Amelda Johnson Sharp, who was born in Ballinger, Runnels County, Texas, in 1888. Although Amelda had to walk a distance to school, she was able to complete the ninth grade. Her husband, Alfred, was the son of Runnels County Treasurer William Frank Sharp. Amelda was said to be such a good seamstress that she could look at a dress in the Sears catalog and make it for her family. They moved to Bristol, Texas, in 1918 where her husband opened a blacksmith shop. Then the family moved to Dallas in 1925 where Amelda attended beauty school and her husband attended barber school. After Amelda and Alfred divorced in 1939, Amelda opened a beauty shop on 12th Street and later in her home at 831 S. Edgefield, in the Oak Cliff area of Dallas. After her son was killed in a car accident in 1943, Amelda retired at the age of 55 and closed the shop. Amelda Johnson Sharp was a successful businesswoman who was able to loan her children funds to begin their businesses.

9171—Chick tea towel motifs, 10c

▲ Listing for Chick day-of-the-week towel embroidery patterns from "Aunt Ellen's Favorite Designs for Gifts, Bazaars and the Home," courtesy of Rose Marie Werner.

▲ Hand towels embroidered with chicks walking in hats, representing Sunday's activity of church attendance, circa 1935, TTU-H2015-053-004-010.

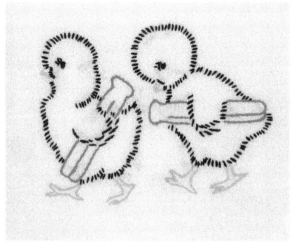

▲ Embroidered hand towels with chicks and clothespins representing Monday's chore of washing, circa 1935, TTU-H2015-053-004-011.

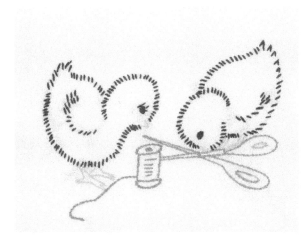

▲ Hand towels embroidered with chicks and thread and scissors, representing Wednesday's chore of mending or sewing clothes, circa 1935, TTU-H2015-053-004-008.

▲ Embroidered hand towels with chicks scrubbing the floor with scrub brush and pail, representing Thursday's chore of cleaning, circa 1935, TTU-H2015-053-004-012.

▲ Hand towels embroidered with chicks creating a market list for Friday's chore of marketing or grocery shopping, circa 1935, TTU-H2015-053-003-013.

▲ Hand towels embroidered with chicks with a rolling pin for Saturday's chore of baking, circa 1935, TTU-H2015-053-003-009.

▲ Hand towels embroidered with Mammy walking to church, representing Sunday's activity of church attendance, circa 1935, TTU-H2015-053-004-078.

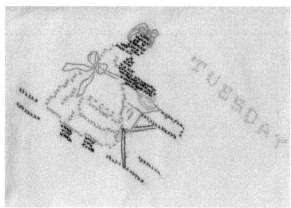

▲ Embroidered hand towels with Mammy figure ironing, representing Tuesday's chore of ironing (so that what was washed on Monday did not mildew), circa 1935, TTU-II2015-053-004-074.

▲ Embroidered hand towels with Mammy figure hand sewing, representing Wednesday's chore of mending, circa 1935, TTU-H2015-053-004-075.

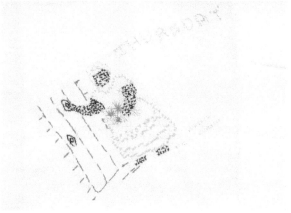

▲ Embroidered hand towels with Mammy figure visiting, probably representing Thursday's task of visiting the sick or visiting your neighbor, circa 1935,

◄ Embroidered hand towels with Mammy figure sweeping. In this set of embroidered towels, Saturday's chore was cleaning, though for some other sets Saturday's chore was baking, circa 1935, TTU-H2015-053-004-077.

The day-of-the-week towels Amelda made are decorated with embroidery and printed cotton feed sack fabric. They were made after 1937, and it is not unreasonable to assume she made them after she closed her shop in 1943, when she likely had more time to do needlework. The day-of-the-week towels Amelda made are of applique in a "Sunbonnet Sue" style with the sunbonnets and skirts made of printed cotton feed sack fabric, and embroidered and finished with a blanket stitch along the edges. The fabric used for these towels is similar to a white sack with colorful woven stripes shown in the previous chapter on White Cotton Sacks: the Hamilton Seed and Coal Co. bag, TTU-H2015-053-003-097 (see p. 20).

▲ Day-of-the-week towel for Sunday, when the activity was attending a church worship service, circa 1940. Gift of Ms. Patricia Hayton, TTU-H2018-007-007.

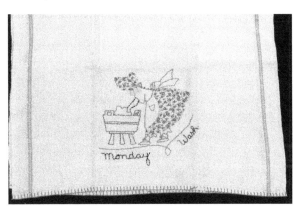

▲ Day-of-the-week towel for Monday, when the main chore was to wash, circa 1940. Gift of Ms. Patricia Hayton, TTU-H2018-007-001.

▲ Day-of-the-week towel for Tuesday, when the main chore was to iron what had been washed the day before so that it did not mildew, circa 1940. Gift of Ms. Patricia Hayton, TTU-H2018-007-002.

▲ Day-of-the-week towel for Wednesday, when the main chore was to mend clothing, circa 1940. Gift of Ms. Patricia Hayton, TTU-H2018-007-003.

▲ Day-of-the-week towel for Thursday, when the main chore in this series was to visit neighbors, circa 1940. Gift of Ms. Patricia Hayton, TTU-H2018-007-004.

▲ Day-of-the-week towel for Friday, when the main chore in this series s was to clean the house, circa 1940. Gift of Ms. Patricia Hayton, TTU-H2018-007-005.

▲ Day-of-the-week towel for Saturday, when the main chore in this series was to bake, circa 1940. Gift of Ms. Patricia Hayton, TTU-H2018-007-006.

◄ Amelda Johnson Sharp (1888-1960), maker of the above appliqued and embroidered day-of-the-week towels, in 1935. Photo courtesy of Pat Hayton.

Manufacturers must have responded to the popular trend, as some bags were printed with motifs that were intended to be embroidered. In the collection are examples of a bag printed for a day-of-the-week towel and two bags printed to be embroidered by "nimble fingers."

◄ Some bags were printed to be embroidered with the days of the week, such as this Percy Kent Bag Company bag, which is designed to be embroidered as a day-of-the-week towel for Friday, circa 1940. After it was embroidered, the bag was to be washed to remove all the printed labels. The instructions printed on the bag state, "Ink easily removed. Don't soak but scrub thoroughly with warm water and soap," TTU-H2015-053-003-556.

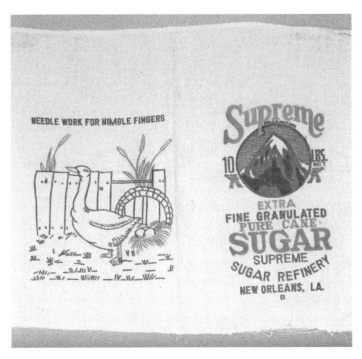

◄ White sugar sack from the Supreme Sugar Refinery, New Orleans, Louisiana, which was printed on the reverse with "Needle Work for Nimble Fingers," and a motif of a duck and its nest in front of a fence, circa 1930, TTU-H2015-053-005-006.

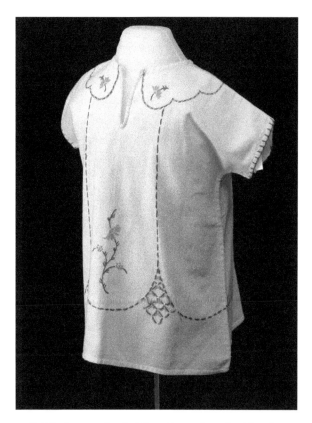

▲ Child's dress made of white cotton sack, embroidered, circa 1930, TTU-H2015-053-001-031.

▲ White sugar sack from the Red Ball Brand, Consolidated Companies, Inc. Plaquemine, Louisiana which was printed on the reverse with "Needle Work for Nimble Fingers," and a motif of a Dutch girl picking tulips, circa 1930, TTU-H2015-053-003-313.

▲ This detail of the Percy Kent printed bag for a luncheon set shows the Dailey's Needlework Saxx logo which states, "My mamma made all these things from Dailey's Needlework Saxx," TTU-H2015-053-005-007, detail.

In addition to the items made for everyday use, the beautiful white fabric also could be embroidered and turned into a child's garment.

White cotton sacks provided a ready source of free fabric, once the logo was removed, for rural households (see Chapter Five for more information on logo removal). The wide variety of decoration applied to this unadorned white fabric documents the large number and variety of households that used the cotton sacks to decorate their homes and clothe their families.

▲ Detail of sack printed to be embroidered into
an apron showing the Needlework Sak logo,
TTU-H2015-053-005-008. Details of the apron printed sack
show that the consumer was to rinse out the bag—so that all
the feed was removed— embroider the piece, and then the
ink could be removed by scrubbing the bag thoroughly with
warm water and soap. Note the typo for "thoroughly."

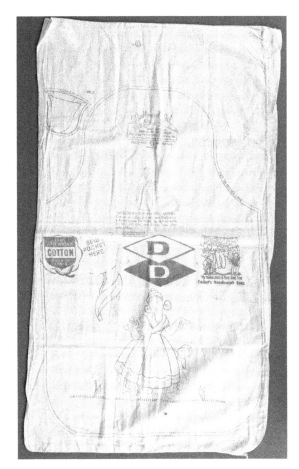

▲ This circa 1930 cotton sack was printed so that it could be
embroidered with a little tulip blossom to be cut and made
into a pocket on the apron. The needlework would cover any
ink left behind after washing. This was called a "Needlework
Sak." Because the apron was printed on just one side of the
cotton sack, it is possible that two aprons could have been
made from one sack. An identical design is printed on a
Tint-sax shown in Chapter One, TTU-H2015-053-005-008.

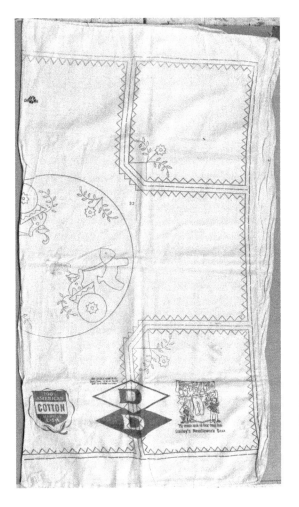

◄ Kent Luncheon Set sack printed on front and back to be
embroidered, circa 1930. Also printed on the sack is, "My
mamma made all these things from Dailey's Needlework
Saxx." The Museum holds examples designed to be made
into an apron, a luncheon set, a pillowcase, and a crib sheet,
TTU-H2015-053-005-007.

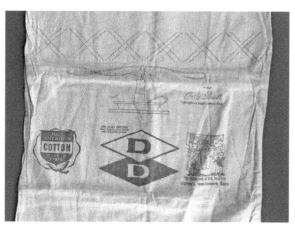

▼ This cotton sack is printed with a border to embroider and make into a crib sheet circa 1930, TTU-H2015-053-005-010.

▲ This cotton sack is printed with a child's size pillow-case to embroider, circa 1930. You can see the size of the child's pillow in the upper right hand corner of this image. Several cases could be made from the fabric in this bag, TTU-H2015-053-005-009.

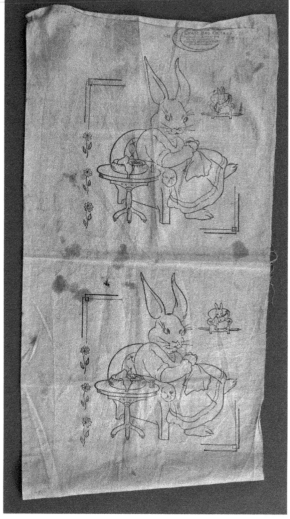

► The Kansas City Chase Bag Company printed sacks such as this with the image of a rabbit sitting in a chair doing needlework to be embroidered. These could be made up into pillows or a quilt. Each sack came with two printed motifs to be embroidered, circa 1930, TTU-H2015-053-005-003.

CHAPTER FOUR ·····COTTON SACKS AND BELGIUM FAMINE RELIEF

Cotton sacks took on the important role of famine relief during World War I in Belgium. In 1914 after being invaded by Germany, the Belgian people were starving as their food was requisitioned by the occupying German military. Cardinal Desire-Joseph Mercier of Belgium, who was then a well-known Catholic cleric, encouraged those who could help to send relief to the starving Belgians. In the United States, the Commission for Relief in Belgium (CRB) was established. Its mission was to provide food relief to war-torn Belgium and was intended to work through voluntary efforts. Herbert Hoover, who would later become President of the United States, led the relief effort. Between 1914 and the end of the war in 1919, almost 11,000,000 Belgians were fed. The CRB coordinated shipments of food past the German submarine blockades and into army-occupied areas and controlled food distribution. Under Hoover's leadership, 697,116,000 pounds of flour were sent, in addition to sugar and grains.

The flour was packaged in white cotton sacks, and these sacks often bore the brands of the American flourmill that had packaged the flour or the foundation that had provided the funds to purchase the flour. Each sack of flour was carefully monitored because the CRB wanted to be sure the flour reached its intended audience and the cotton sack didn't fall into German hands. The cotton fabric in the sack could be used for ammunition and the flour could be stolen and diluted with poor quality flour. So a careful watch was kept on the sacks, both full and empty.

Empty flour sacks were carefully accounted for and distributed to professional schools, sewing workrooms, convents, and individual artists. The professional schools in Belgium that received the bags were separate from trade schools and specialized in training girls to sew, embroider, and make lace. The sewing workrooms were large centers that were established in the major Belgian cities during the war. These workrooms were intended to provide work for thousands of unemployed Belgians. Girls and women made Belgian lace, embroidered textiles, and repaired and remade clothing. The fabric from the flour sacks was used to make new clothing, accessories, pillows, bags, and other functional items.

▲ Image of a storehouse for the American Commission for Relief in Belgium (CRB).

▲ Antwerp Music Hall during the German occupation of Belgium, where women assembled to sew, embroider, make lace, and embellish the flour sacks. Source: Charlotte Kellogg, *Women of Belgium; Turning Tragedy to Triumph* (New York and London: Funk & Wagnalls Company, 1917).

Many different types of Belgian acknowledgments were created in response to these gifts. Edward Eyre Hunt, who ran the CRB operations in Antwerp, recorded these activities in his 1916 book, *War Bread: A Personal Narrative of the War and Relief in Belgium*. Hunt states that the offices became "stacked with beautiful souvenirs for the American people including silk banners, wrought metal boxes, leather work, a magnificent carpet woven on the famous looms at Westerloo and intended for the White House, lace souvenirs of great value and rare beauty, etc."[1] Hunt goes on to say,

> But the most touching and most original souvenirs were made of American flour sacks. No one knows who first planned these gifts. They seemed to spring up spontaneously in all parts of Belgium as the simplest expression of the feelings of the people. To take the sacks, emptied of their precious flour, and turn them into souvenirs for the American donors was an inspiration, and some of the results have been very beautiful. Most of them are embroidered with designs in finest needlework, and lettered 'Homage to America,' 'Thanks to America,' 'Out of Gratitude to America,' 'grateful Belgium to Kind America,' 'To the Saviour[sic] of Belgium,' or in simplest Flemish or French, 'Thanks.' One of them shows Lady Columbia with a Belgian baby in her lap and is inscribed, 'The Protecting Mother of Belgium.'
>
> For to the Belgian people one thing seems very clear: that they would have starved without the intervention of America.[2]

Hunt and Charlotte Kellogg, the only American woman working with the CRB in Belgium, both reported in their books that in Antwerp these bags were being embroidered in the Antwerp Music Hall, which had been converted into a sewing room to create garments (and jobs) for those displaced by the war. Kellogg reported,

> In one whole section the girls do nothing but embroider our American flour sacks. Artists draw designs to represent the gratitude of Belgium to the United States. The one on the easel as we passed through, represented the lion and the cock of Belgium guarding the crown of the king, while the sun—the great American eagle rises in the East. The sacks that are not sent to America as gifts are sold in Belgium as souvenirs. Each sack has its value before being worked. Many of them—especially in the north of France—have been made into men's shirts, and tiny babies' shirts and slips.[3]

Kellogg is the only one to mention that the sacks not sent as gifts to America were sold in Belgium as souvenirs, but being an eyewitness her report has authenticity. Despite the desperate situation for most of the Belgians, there were some who still had economic means and were accordingly charged something for the food the CRB was distributing. These funds were used to pay for goods and services for which the CRB had to pay.[4]

The February 1917 issue of the magazine *The Modern Priscilla* reported on the phenomenon of Belgium women embroidering the flour bags as thank-yous for the famine relief:

> To-day the women of Belgium are expressing their gratitude.
>
> American flour goes to Belgium in bags, and the domestic uses to which a flour-bag can be put are known best to the woman who has been in the extremity of need. The bags help to clothe the family, serve as towels and cover the children's beds.
>
> The love and gratitude poured out upon the American volunteers, who have fed the hungry, clothed the naked and carried on the thousand and one acts of relief and encouragement that can only be performed by men who love their fellows, has been touchingly and beautifully stitched into these

precious relief flour-bags. The idea first occurred to an Antwerp woman who embroidered on the unprinted side of a bag a bouquet of heather and the words "Uit Daukbourbrid –with gratitude," stuffed the bag with feathers from her last fowl and presented the pillow to the young American who was the visible manifestation of care and understanding for shorn Belgians in her province. The notion spread like wildfire among the women, not of Antwerp alone, but of all Belgium. 'Seeds of Kindness grown into Flowers of Love,' the motto on one of the pillows expresses the sentiment of the whole flour-bag souvenir movement.[5]

The Herbert Hoover Presidential Library and Museum holds many of the embroidered flour sacks from Belgium in their collections because so many were given to Hoover due to his leadership of the relief effort. However, one example of these relief sacks came into the Museum of Texas Tech University's collection through the Nickols collection.

The sack documents the work of The Rockefeller Foundation. The newly established Rockefeller Foundation began in 1914 to open its coffers to a variety of war-related causes that required immediate attention.[6] The Rockefeller Foundation contributed almost a million dollars (in 1914 dollars) to the work of Belgian relief.[7] In addition to money for food as well as funds to pay those who loaded and unloaded the ships, the Rockefeller Foundation supplied generous gifts of clothing so important to the relief effort.[8]

While Germany occupied Belgium, the Belgian flag was not permitted to be flown, nor were any representations permitted.[9] However since the German occupiers were not the intended recipients of the flour sacks, many of these thank-yous featured Belgium's flag or were embroidered in the black, red, and yellow colors of Belgium.

It is likely that more of these embroidered thank-you sacks exist undocumented around the world. Textile historian Annelien van Kempen is compiling a list of institutions that have flour sacks embroidered by the Belgium women in their collections. However, this resource is only available in Dutch.[10] The Museum of Texas Tech University is fortunate to have one of these treasured thank-yous in the collection.

◄ Article from *The Modern Priscilla* that discusses the Belgian women embroidering flour sacks as thank-yous to the United States. Source: Clothing and Textiles Research Library, Museum of Texas Tech University.

▲ This flour sack, originally used to send flour to Belgium for famine relief during World War I, was embroidered over the inked label. The back of the sack was embroidered with interlocking Belgian and United States flags, representing the friendship between the two countries. The edge around the flags is a ribbon. This sack documents the work of The Rockefeller Foundation in the Belgian relief effort and is embroidered with the years 1914, 1915, 1916, and 1917, TTU-H2015-053-005-001.

▲ All 21 instructional booklets in the Nickols Collection of Printed Cotton Sacks at the Museum of Texas Tech University.

CHAPTER FIVE ⸺ INSTRUCTIONAL BOOKLETS SPREAD FASHION AND QUILTING INFORMATION

In an effort to promote the use of cotton bags as free, reusable fabric, instructional booklets were printed from about 1924 until about 1970 by entities such as the Household Science Institute, the Textile Bag Manufacturers Association, and the National Cotton Council.[1] This booklet program was intended to offset the price disadvantage cotton bags faced over paper by increasing consumer demand for cotton bags. The instructional booklets are a source of fashion and quilting information and document changes in how consumers purchased sewing patterns and the changes in the production of cotton sacks. The booklets encouraged the use of the fabric for clothing for oneself and the family as well as making bazaar items, home decorating accessories, toys, and quilts. A total of 21 instructional booklets held in the Nickols Collection at the Museum of Texas Tech University form the basis of this chapter. These primary documents are supplemented by contemporary publications, advertisements, secondary sources, and correspondence with National Cotton Council (NCC) and Simplicity Patterns staff.[2] Al-most all the instructional booklets are undated, but research has enabled accurate dates to be provided for them.

As mentioned in Chapter One, cotton sacks, which packaged staples and feed, were a source of free fabric for women from the end of the nineteenth century through the 1960s.[3] Although the cotton sacks were used for a wide variety of staples, the manufacturers particularly designed sacks for human food and chicken feed with women's interests in mind. Women were the prime food preparers and were therefore most involved in the selection of the flour, sugar, and other staples households would purchase. Women also typically supplemented their family income through the sale of eggs and produce[4] that they raised, so they were also involved in the purchase of chicken feed. Rural women were the first targeted by the publishers to receive these free booklets, but later urban women received them as well.

The Household Science Institute was the first to publish this type of instructional booklet with *Sew-*

ing with Flour Bags. The booklet announced its purpose in the first line of this mid 1920s publication, stating:

> . . . not all empty flour bags end their lives as humble dish towels. Many of them find their way to the sewing room to hobnob democratically with silks and satins and laces. From there they emerge in a variety of forms so practical in use and attractive in appearance that their lowly origins would never be suspected.

Although not dated, the booklet was likely published during the presidency of Calvin Coolidge, which began on August 2, 1923 and ran until March 4, 1929, because it features pajamas made for President Coolidge, and text referencing the "ladies of the Millard Avenue, Chicago, Presbyterian Church [who] were anxious to do something to show their approval and appreciation of President Coolidge's program of thrift and economy. . . . A warm letter of appreciation from the White House was sent to the ladies, voicing Mr. Coolidge's acceptance and appreciation of the unique gift."[5] Based on the dates of Coolidge's presidential term, the booklet would likely date 1924–1928.

Articles in *Sewing with Flour Bags* included basic information about cotton sacks as well as images of the variety of items that could be made from the bags. The format established by this initial instructional booklet was similar to that followed by subsequent publishers. The Textile Bag Manufacturers Association published instructional booklets from circa 1930–1941. Their work was taken over by the National Cotton Council, which was launched on November 21, 1938. The National Cotton Council published the instructional booklets as part of a campaign that was "designed to hold and expand the estimated 425,000-bale market of cotton bags against the growing threat of paper."[6] They worked with the Textile Bag Manufacturers Association to use the booklet to emphasize "cotton's re-use value and provided sewing suggestions for the thrifty."[7]

▲ Instructional booklet, *Sewing with Flour Bags*, published by The Household Science Institute circa 1926; patterns were available through the pattern department of the newspaper. This booklet likely dates to 1924-1928 and is the earliest example in the Nickols Cotton Sack Research Collection, TTU-H 2015-053-009-001.

The basic information published in the instructional booklets included a schedule of sizes listing the approximate size of fabric in various bags, how to open a bag to make it flat, and how to remove the inked logo. The variety of items, for which drawings were included and patterns available by mail order, in the first booklet included a beach coat, aprons and dresses for housework, and pajamas. A section on children's clothing stated that "using flour bags is not only a[n] economical way of teaching the children to sew, but it furnishes at the same time a practical, early lesson in thrift and conservation." Also in the booklet were instructions on making "tasteful luncheon sets," card table covers, bazaar items, home accessories to organize closets and keep out dust, gifts, home decoration items, baby gifts, toys, and quilts. As an aid to home dressmakers, women were encouraged to make up their garments first in flour sack fabric before cutting into expensive dress fabric.

So that consumers would know how much fabric each bag provided, a chart was included in the early instructional booklets with measurements of each type of bag in inches. In fact, according to Pat Nickols's measurements of a variety of 100-pound bags, these bags did vary in size even for bags of the same commodity and product weight.[8] The following chart should be considered the average for the commodity and weight:

▶ Instructional booklet, *Sewing with Cotton Bags*, published by the Textile Bag Manufacturers Association between 1930 and 1936 with mail-in patterns from The Beauty Pattern Company, TTU-H2015-053-009-003.

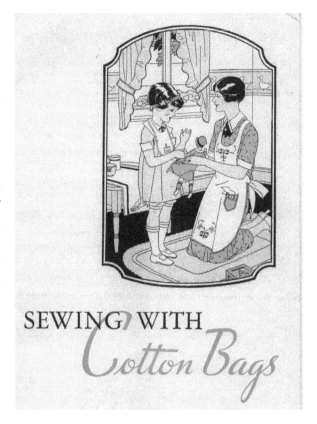

SIZES	FLOUR BAGS	SUGAR BAGS	SALT BAGS	MEAL BAGS	FEED BAGS
5 lb.	15 x 19	13 x 16	13 x 14	15 x 16	40 x 46
10 lb.	18 x 23	16 x 21	16 x 17	18 x 22	40 x 48
25 lb.	26 x 26	22 x 27	18 x 26	26 x 27	40 x 50
50 lb.	30 x 34	—	—	—	40 x 52
100 lb.	36 x 42	36 x 40	30 x 36	36 x 44	40 x 54

The center spread of both *Sewing with Flour Bags* published by the Household Science Institute and *Sewing with Cotton Bags* provided ideas for items that could be made for charity bazaars. Both recommended making refrigerator bags for the storage of vegetables. A set of feed white feed sack vegetable bags with pretty red embroidery is part of the museum's collection (see p. 55).

Removing ink from the bag so that only white or cream-colored fabric remained was one of the booklet topics. The cotton bag manufacturers want-ed consumers to think the ink removal was an easy process, but apparently it was difficult, as the many remaining faded printed logos on feed sack objects attest.[9] Tracking the changes in these directions helps to date the undated booklets. The Household Science Institute booklet advised the following:

The ease with which stamping may be removed from flour bags depends on the kind of ink that has been used. Under ordinary circumstances it is sufficient to cover the inked places with lard or soak them in kerosene overnight. Then wash out in lukewarm water. If only a faint, barely discernible marking may still be seen, it is safe to assume a few washings will remove the remaining traces.[10]

In addition to including the directions above, the revised edition published by the Textile Bag Manufacturer's Association in 1937 also included another method:

Wet a bar of laundry soap and rub on the dry bag. Repeat wetting the soap and rubbing on until the bag is covered with a thick layer of soap. Roll the bag up and let it stand for several hours before washing. When the bag is washed and boiled, it will be as white as any muslin.[11]

Bag Magic published circa 1944 gave advice about using the cotton sack's string under the heading, "Save the String, Too:"

The string used in stitching a cotton bag . . . can make many pretty edgings and even whole doilies and table mats. The bag is sewn with a chain stitch that usually starts in the lower corner near the fold. By cutting the chain close to the bag and lower threads at this point and pulling both, the stitching is ripped in a jiffy. Because the thread is of fine cotton, you can tint it along with the bag fabric should you prefer to work in color. Rip bag before tinting!

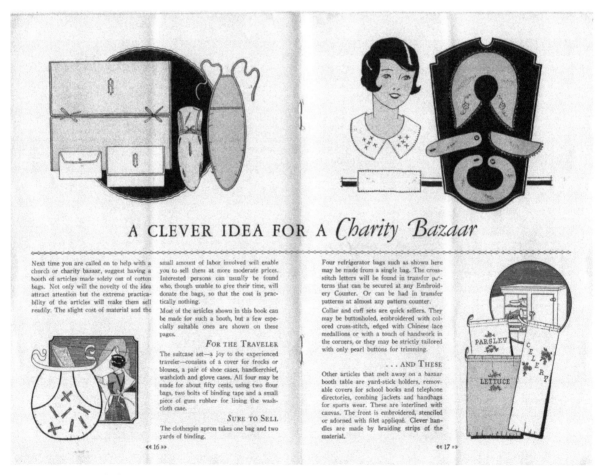

▲ This center spread of *Sewing With Cotton Bags*, published by the Textile Bag Manufacturers Association between 1930 and 1936, provided ideas for items, such as vegetable bags, that could be made and sold for charity, TTU-H2015-053-009-003.

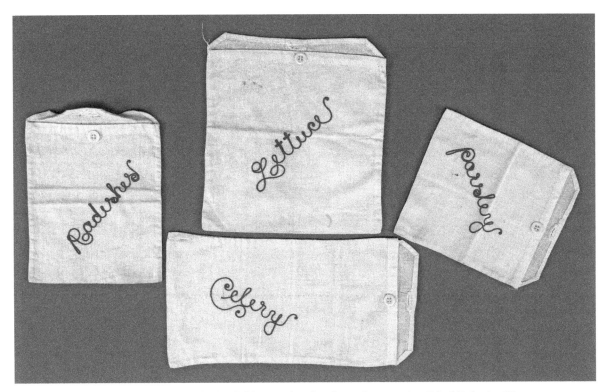

▲ Refrigerator storage bags made from feed sack fabric, embroidered in red, circa 1935. Gift of Catherine Watts, TTU-H1973-262-004a-d.

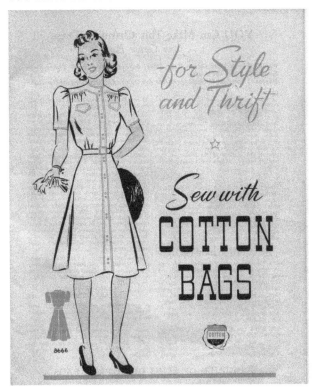

◄ Instructional booklet, *For Style and Thrift Sew with Cotton Bags*, published by the Textile Bag Manufacturers Association, revised January 1941 with mail-in patterns from The Beauty Pattern Company, TTU-H2015-053-009-005.

◄ Instructional booklet, *A Bag of Tricks for Home Sewing*, published by Sears Roebuck and Company in cooperation with the National Cotton Council of America between 1942 and 1943, with mail-in patterns from Famous Features Syndicate, TTU-H2015-053-009-007.

▼ Information on how to use the string that could be saved when the bags were taken apart and suggestion for gifts, featured in the instructional booklet, *A Bag of Tricks for Home Sewing*, published by Sears Roebuck and Company in cooperation with the National Cotton Council of America between 1942 and 1943, TTU-H2015-053-009-007.

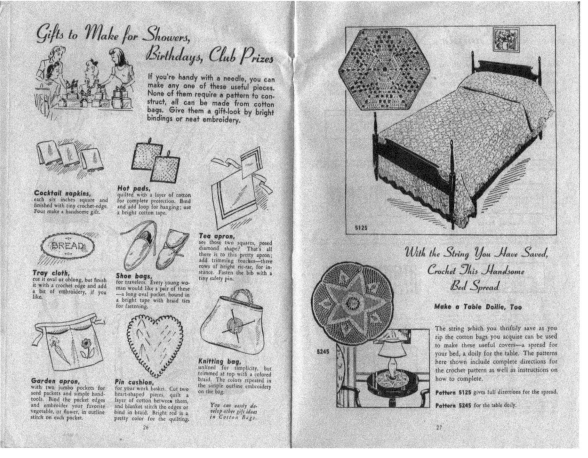

◄ Instructional booklet, *Kasco Home Journal*, published by Kasco in 1943 with mail-in patterns from Fashion Headquarters in New York City. A subtitle for the booklet, "Sew for Victory with the Lovely New Dresprint Fabrics in Kasco Sax," indicates that this booklet focused on a particular brand of printed cotton sacks, the Kasco Sax. All other booklets discussed in this chapter were generic in nature, encouraging the home sewer to use the cotton sacks without regard to brand, TTU-H2015-053-009-006.

▲ Instructional booklet, *Bag Magic*, published by the National Cotton Council in 1944 with mail-in patterns from Famous Features Syndicate, TTU-H2015-053-009-008.

By the early 1940s, the textile bag manufacturers had been able to improve the inks so that they were easier to remove, and the directions provided in the next editions by the National Cotton Council eliminated all reference to lard or kerosene as cleaning agents:

> Soak the bag overnight in heavy soap suds, then wash thoroughly in warm suds, gently rubbing the parts covered by the ink. If all traces of ink are not then removed, boil in soapsuds for at least ten minutes. Be sure this is done before cutting the bag or tinting it![12]

This statement documents that cotton sacks were dyed before they were used. The back of the Devil's Highway quilt, shown on p. 88 in Chapter 7 on feed

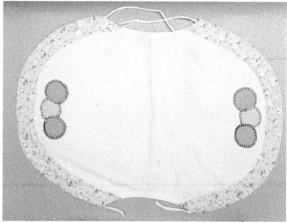

◄ Instructional booklet, *A Bag of Tricks for Home Sewing*, published by the National Cotton Council in 1945 with mail-in patterns from Famous Features Syndicate, TTU-H2015-053-009-009.

▲ Made from white feed sack fabric and trimmed with printed fabric that forms a casing around the outside, this was a cover for a casserole dish and is similar to the Pot Luck Cozie shown in the 1945 publication, *A Bag of Tricks for Home Sewing*. Gift of Lenna DeMarco, TTU-H2018-114-017.

sack quilts was made from a flour sack that had been home-dyed blue.

By the time *Thrifty Thrills with Cotton Bags* and *How to Sew and Save with Cotton Bags* were published by the National Cotton Council, probably in 1946, paper labels were in use by some bag manufacturers and the directions for ink removal indicated that the current inks were easier to remove:

> Removing the printing on cotton bags has been so greatly simplified by use of new wash-out inks that it is now no trouble at all. Simply soak the bag in warm soapy water. The new inks need little coaxing to do a quick disappearing act. Brand names are sometimes printed on paper labels glued to the bags. Only a few seconds after dipping the bag in water, you can easily strip the label from the sack. Inks and labels should be removed before tinting or cutting.

The National Cotton Council partnered with Simplicity in the mid-1940s, and their first booklet together bears the name of Mary Lynch, a Simplicity employee. Her work dates (1944–1948) and her marriage date (September 1948) in turn help date the booklet. The two booklets bearing her name are the only two among the 19 examples of instructional booklets in the Nickols Collection that have an author's name. Based on her maiden name in one issue and her married name in the other, a date can be assigned to the two publications. The booklet bearing only her maiden name must be 1944–1947, but, because of the way the label removal directions are written in *Thrifty Thrills*, the date narrows to 1947. The label removal directions for *Smart Sewing with Cotton Bags* with Mary Lynch's byline stated, "Wash out printed labels in warm, soapy water, soaking overnight if you wish. They are printed with removable colors. Some bags now have paper labels—even

Unrationed Hospitality

Inviting friends in for the afternoon or evening can be something you do graciously, naturally and without long planning if you have at hand the hostess-aids that you can make easily from cotton bags. Below are two practical suggestions.

Pattern 5141: a bridge table cover and chair backs that make your invitation to "come in for a game" something of an occasion. Pattern includes cut-outs for elaborate applique and embroidery to trim. You use the larger bags to make the covers, cutting them to the size of your table and chair.

Pattern 5121: a charming tea cloth and napkins for the thoughtful hostess, each corner of the cloth slashed to hold the napkin firmly in place inside the applique tea pot. Pattern includes directions for making and outline pattern for applique cut-out.

5141

"Pot Luck" Cozie

Mrs. Fred M. Hensley, Zurich, Kansas submitted this prize-winning idea. Cover slips over casserole. Several covers can be made from one medium-size bag if remnant print is used for contrast. Bias tape trims seam and drawstring is made from twisted string.

5121

14

▲ Image of "Pot Luck" Cozie from the instructional booklet, *A Bag of Tricks for Home Sewing*, published by the National Cotton Council, 1945, TTU-H2015-053-009-009.

▲ Instructional booklet, *How to Sew and Save with Cotton Bags*, published by the National Cotton Council with patterns available by mail through Famous Features Syndicate, circa 1946, TTU-H2015-053-009-012.

▶ Instructional booklet, *Thrifty Thrills with Cotton Bags*, published by the National Cotton Council with patterns available by mail through Famous Features Syndicate, circa 1946, TTU-H2015-053-009-011.

◄ Instructional booklet, *Smart Sewing with Cotton Bags*, by Mary Lynch, published by the National Cotton Council with patterns by Simplicity available (for the first time) at a local store, circa 1947 TTU-H2015-053-009-013.

▼ Instructional booklet, *Sew Easy with Cotton Bags*, by Mary Lynch Lincoln, published by the National Cotton Council with Simplicity patterns available at a local store, 1948 TTU-H2015-053-009-014.

◄ Mary Lynch prior to her marriage to Edmond Lincoln on September 3, 1948. Photograph by Bradford Bachrach, an important photographer of society women and female businesswomen. Courtesy of her children, Edmond Lincoln and Dr. Mary A. Lincoln.

◄ Instructional booklet, *Needle Magic with Cotton Bags*, published by the National Cotton Council with Simplicity patterns available at a local store, 1949–1950. An identical booklet in the collection was published by Dominion Textile Company, Limited, Montreal Quebec, TTU-H2015-053-009-015.

► Instructional booklet, 1952 *Pattern Service for Sewing with Cotton Bags*, published by the National Cotton Council with Simplicity patterns available at a local store, 1952, TTU-H2015-053-009-017.

▲ Instructional booklet, *1953 Pattern Service for Sewing with Cotton Bags*, published by the National Cotton Council with Simplicity patterns available at a local store, 1953, TTU-H2015-053-009-018.

easier to remove."[13] The same directions are printed under her married byline, Mary Lynch Lincoln, in 1948.[14]

Needle Magic with Cotton Bags, which seems to date to 1949–1950, provided no information on removing ink or labels and Mary Lynch Lincoln's name does not appear on the booklet. The *1952 Pattern Service for Sewing with Cotton Bags* advised the following:

> Remove the Label—Almost all bags have band labels for brand identification. Soak the bag in water and the label comes off in a jiffy! Some brand names are printed in washable inks that come out easily when soaked in warm soapy water.[15]

The same advice was included in the National Cotton Council's *1953 Pattern Service for Sewing with Cotton Bags*,[16] *1954 Idea Book for Sewing with Cotton Bags*,[17] and the circa 1955 *Ideas for Sewing with Cotton Bags*.[18] *Ideas for Cotton Bag Sewing, 1963* did not include any directions for preparing the cotton bags. This is not surprising as, by 1953, one of the leading bag manufacturers, the Fulton Bag and Cotton Mills, exclusively used paper labels that were easy to remove, and it is likely that all other bag manufacturers were doing likewise.[19]

Another way to document the dates on the instructional booklets is to consider the sources of the patterns in the booklet. At the beginning of the twentieth century McCall's and other women's magazines were the source of patterns, but consumers had to mail in for the patterns they wanted. The early editions of the cotton bag instructional booklets followed this system with consumers having to mail in their request, with accompanying payment, for patterns first to the Patterns Department of the Newspaper, then in the 1930s under the Textile Bag Manufacturer's Association, to The Beauty Pattern Company in New York City.

The first editions of the booklets published by the National Cotton Council from 1942–1946 included patterns that were available through Famous Features Syndicate also in New York City. *Bag Magic for Home Sewing*, published circa 1944, stated that "the National Cotton Council has developed a family wardrobe from cotton bags which will be shown throughout the country. Some of these garments are featured in this book together with the pattern numbers from which they are made."

The National Cotton Council first partnered with Simplicity Patterns in 1947 when Mary Lynch worked on the booklets. This was a significant change because Simplicity pioneered placing patterns for sale in a local store so that the consumer didn't have

YOUR PFAFF DEALER PRESENTS
1954 YOUR—
Idea Book
for sewing with cotton bags

WIN THE
FABULOUS

PFAFF
DIAL-A-STITCH SEWING MACHINE.
SEE CENTER SPREAD FOR
CONTEST DETAILS

▲ Instructional booklet, *Your Idea Book for Sewing with Cotton Bags*, published by Pfaff Sewing Machines with the National Cotton Council with Simplicity patterns available at a local store, 1954, TTU-H2015-053-009-019.

students, but she was also a published author of magazine articles,[22] making her the perfect candidate to work with the National Cotton Council to revise the instructional booklets.

Until 1954 all the instructional booklets were illustrated with line drawings of the garments. Beginning in 1954, black and white photographs illustrate the covers, but line drawings of the garments continued to be used in the interior of the booklet. This same format continued to be used for the booklets published in 1955 and 1963.

The 1954 and 1955 booklets both describe "Save with Cotton Bags" Sewing Contests, which were held at regional and state fairs, and the booklets include the contest rules. Prizes included a variety of home appliances, such as a two-oven Universal gas range, sewing machines, and small household items like Betty Crocker Cook Books from General Mills, Inc.[23] By 1963 this contest had likely been phased out as the information is not included in the instructional booklet published that year. The written history of the National Cotton Council of America describes how the program continued:

> Later in 1969–70, when cotton bags were cotton's largest single market, 'appropriate promotional emphasis was provided.' Virgene Robinson had operational responsibility for the campaign and, working with McCall Patterns, developed 50 sets of traveling wardrobes for fashion shows, which were scheduled across the country year after year. Accompanying the wardrobes were a million cotton bag wardrobe booklets made available to home sewers."[24]

to mail in for them.[20] Simplicity first placed their patterns for sale at Woolworths and then other stores. Partnering with Simplicity likely meant that The National Cotton Council didn't have to come up with its own fashion designs and that Simplicity assisted with publication costs as part of their marketing budget.

Once Simplicity became a partner, the format of the booklet was revised, although most of the same topics were covered. Mary Lynch had been a Home Economics teacher in Pennsylvania before moving in 1944 to New York City to work in Simplicity Pattern's education department.[21] Mary Lynch was particularly interested in teaching sewing to high school

Unfortunately the Museum's collection does not yet have a copy of this 1969–70 booklet produced under the supervision of Virgene Robinson, but it may be the booklet shown on p. 65, *Cinderella Fashions from Cotton Bags.*

The earliest versions of the instructional booklets all had drawings of garments made of solid col-

◄ Instructional booklet, *Ideas for Sewing with Cotton Bags*, published by the National Cotton Council with Simplicity patterns available at a local store, 1955, TTU-H2015-053-009-020.

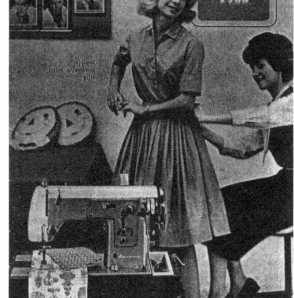

► Instructional booklet, *Ideas for Cotton Bag Sewing*, published by the National Cotton Council with Simplicity and Butterick patterns available at a local store, 1963, TTU-H2015-053-009-021.

▲ Instructional booklet, *Cinderella Fashions from Cotton Bags*, published by the National Cotton Council with McCall's patterns available at a local store, 1969–1970, image courtesy of Merikay Waldvogel.

instructional booklet likely was part of the new campaign to increase the use of cotton sacks by printing them in beautiful prints. The bag manufacturer, the Percy Kent Bag Company, first put prints on bags in 1937, but the use of printed fabric for cotton bags did not become widespread until 1940–42.[26]

▲ The instructional booklet, *A Bag of Tricks for Home Sewing*, published circa 1942–1943, was the first booklet published by the National Cotton Council of America to show a printed fabric, likely from a cotton sack.

ored fabric, or embellished with a small amount of printed fabric, which was likely not from a cotton sack but was rather purchased fabric. Instructions in those booklets also offered directions on how to dye to white cotton fabric because cotton sacks with pretty prints were not available until the late 1930s. The 1943 Kasco *Home Journal; Sew for Victory with The Lovely New Dresprint Fabrics in Kasco Sax,* is the first instructional booklet to feature printed fabrics throughout the booklet. Kasco feeds were sold only east of Indiana.[25] Clearly designed to get women to purchase Kasco Feeds, many fashionable outfits are illustrated. These were available from Fashion Headquarters on 7th Avenue in New York City. This

As noted earlier, this initial campaign to use cotton sacks for clothing was directed at rural women whose households regularly consumed large quantities of flour and feed, but later campaigns aimed at urban women told them where to acquire the bags in town. In 1944 the *New York Times* reported the booklet *A Bagful of Tricks*, was being published by the National Cotton Council. As the *New York Times* did comment, "Even the most enthusiastic innovator, however, might wonder how the average city housewife who does not lug home her flour or sugar in fifty pound bags, will be able to cooperate."[27] Of particular interest is that an apron in this booklet shows the use of printed fabric combined with solid (see p. 65). This image is the first representation of printed fabric in the instructional booklets published by the National Cotton Council.[28]

Although Simplicity has reused numbers on its pattern envelopes, the graphic design of the envelope helps date a particular pattern in the early years. During the 1940s, the envelope font was printed in a script style.[29] In 1944 the word "printed" appeared in red on the envelope and, by the end of that year, the pattern price had increased to 25 cents. A double black bar above the price was added to the pattern envelope design in 1951. The website Cemetarian Hook Purl & Thimble provides dates for vintage Simplicity Pattern numbers. The pattern #3412 pictured here has the script style of the lettering and, according to Cemetarian Hook Purl & Thimble, was available in 1950.[30] This supports the date of 1949–50 for the booklet *Needle Magic with Cotton Bags*, on page 61 in which it is featured and documents that the latest Simplicity patterns were included in the instructional booklets.

The instructional booklets' availability was also advertised on the cotton sack itself. The page 67 bag of Nutrena Feeds from the Nutrena Mills, Inc., a division of Cargill, Inc. of Minneapolis, Minnesota, was printed with information about how to obtain a copy of the instructional booklet *Needle Magic with Cotton Bags*.

▲ Envelope for Simplicity Pattern #3412 which appeared on page 9 of the instructional booklet, *Needle Magic with Cotton Bags*, circa 1949-1950.

Time Magazine in 1949 discussed the ongoing promotional booklets, which were intended to expand the use of cotton sacks by the cotton industry over the paper market.[31] Rural Americans had used the cotton bags for years, but after WWII the papermakers edged higher-pried cotton out of the bag market, and the National Cotton Council "decided to go after city folks too."[32] The bag manufacturers upgraded the fabric designs and with this shift came its partnership with nationally known Simplicity patterns. The Bemis Brothers Bag Company ad for 1947 proclaimed, "Women go for these New York Fashions! Straight and fast into Bemis plants go these exclusive prints from the world's fashion center."[33] Other cotton bag manufacturers also worked to come up with high fashion prints by hiring New York designers. Bemis pre-tested its fabrics by having women select the ten best designs in order of preference. The sample designs with the most votes were called "Panel-Picked Patterns." These survey results were the basis of what the New York designers then worked with to design the upcoming line.[34]

The *Dallas Morning News* recorded the use of feed sacks along with old ribbons and flowers by an art student doing a take-off on new French fashions in 1958.[35] Despite these beautiful fabrics, with the improving economy and with a larger percentage of Americans living in urban rather than rural areas, making clothing, quilts, and other items out of feed sacks would eventually decline after the 1960s.

► This Nutrena Feeds sack mentions the *Needle Magic with Cotton Bags* instructional booklet on the bag, circa 1949, TTU-H2015-053-003-144.

◄ Detail of Nutrena Feeds sack mentioning the *Needle Magic with Cotton Bags* instructional booklet, TTU-H2015-053-003-144.

◄ Clyde Mitchell Davis Livingston (Mrs. Connelly E. 1903-1987) of Atlanta, Texas, in a dress she made from printed feed sacks, circa 1945. Photo courtesy of Mari L. Madison.

► Ann Livingston, later Carvalho (1937-2015), wearing a feed-sack fabric vest and skirt made by her mother, Clyde Davis Livingston (Mrs. Connelly E.) in Atlanta, Texas, circa 1945. Photo courtesy of Mari L. Madison.

CHAPTER SIX ⁓ FEED SACK GARMENTS

Cotton sacks were considered free fabric, therefore it is not surprising that many rural households would cut them up and recycle them into clothing for men, women, and children. Indeed the creation of the instructional booklets described in the previous chapter went a long way towards popularizing this practice. Garments of both outerwear and underwear were made from cotton sacks, more so after the advent of the pretty printed sacks in 1937. The Museum holds several examples of feed sack garments from both the Nickols collection as well as that of quilt historian Lenna DeMarco and other individuals.

Underwear was something women didn't mind making from cotton sacks, even before the printing was easy to get off. A 1921 issue of *American Cookery*, the Boston Cooking School magazine, outlined the value of using these white cotton sacks:

> In homes where little folks are growing up, not a scrap of sacking need be wasted for each sack takes the place of an equal quantity of muslin, since there are so many necessary little garments to be made. The sacking, while not fine in quality, is most serviceable for drawers, petticoats, underwaists, etc.[1]

It was not expected that anyone outside the home would see undergarments, so the printing on one didn't matter (see p. 70). *Time* magazine reported that a manager from Pillsbury Flour was quoted as memorably saying, "They used to say that when the wind blew across the South you could see our trade name on all the girls' underpants."[2]

The most numerous feed sack garments in the Museum's collection, at 13 examples, are the aprons. Since aprons were intended to catch spills and food stains, it is not surprising that they were made from the free fabric in cotton sacks. This cheaper fabric was used to protect possibly finer dress fabric. These aprons were also given as gifts. In 1950 among the 200 US delegates to the 6th triennial global conference of the Associated Countrywomen of the World, a Colorado woman, Mrs. Platt Craig packed more than 400 attractive aprons made from feed sacks. In each of these aprons' pockets was a letter from the

◄ Slip made from feed sacks with printing on front and back. The label reads "For export Net weight 100 lbs. Gross weight 100.45 lbs. Wheat flour 72% extraction Enriched Bartlett & Company, Grain General Offices Kansas City, Missouri U.S.A. Milled in U.S. A. For Jamaica PA-JM-7004", TTU-H2015-053-001-015.

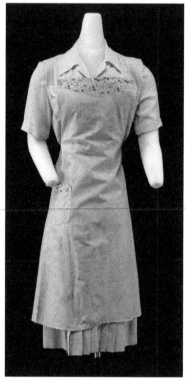

◄ Apron made from white bleached flour sack fabric from the Washburn Crosby Co. bearing the brand, "Kansak," with ink not successfully removed, and trimmed with purchased contrasting fabric that is not feed sack fabric, circa 1927. Gift of Lenna DeMarco, TTU-H2018-114-009.

◄ These pages from the circa 1926-1927 instructional booklet, *Sewing with Flour Bags*, document the apron shown above as #5882, Ladies Apron, that required one flour bag and 1/8 yard of 36-inch-wide contrasting fabric for the yoke and pockets.

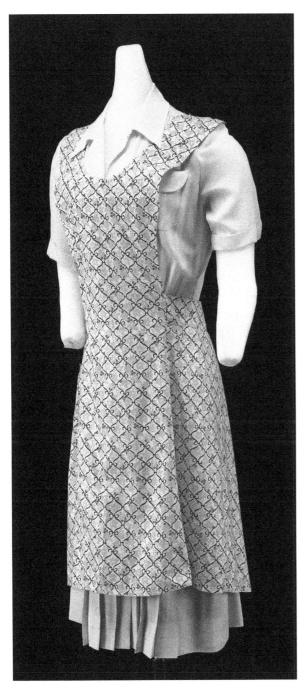

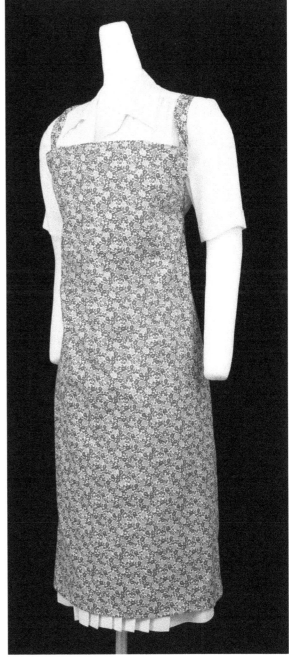

◄ Full apron with straps that crisscross in the back of printed cotton sack fabric, circa 1940. Gift of Lenna DeMarco, TTU-H2018-114-010.

► Full apron made from feed sack fabric, circa 1940, TTU-H2015-053-001-001

◄ Full apron made from feed sack fabric with a heart-shaped bodice and blue rick-rack decoration, circa 1940, TTU-H2015-053-001-003.

► Full apron made from feed sack fabric trimmed with green bias tape along edges, circa 1940, TTU-H2015-053-001-004.

◄ Full apron of printed cotton sack fabric blocks arranged in a "Trip Around the World" pattern, trimmed with purchased brown bias binding, circa 1940. Gift of Lenna DeMarco, TTU-H2018-114-014.

► Half apron made from feed sack fabric with solid yellow accents of purchased fabric, circa 1940, TTU-H2015-053-001-008.

◄ Half apron made from feed sack fabric pieced in a Dresden Plate pattern, circa 1940, TTU-H2015-053-001-009.

► Half apron made from pieced feed sack fabric, circa 1940, TTU-H2015-053-001-013.

◄ Half apron made with similar block arrangement of alternating white and blue printed feed sack fabric, circa 1940. Gift of Lenna DeMarco, TTU-H2018-114-015.

► Half apron of pieced blocks of printed cotton sack fabric edged with purchased blue bias binding, circa 1940. Gift of Lenna DeMarco, TTU-H2018-114-013.

woman who sewed the apron, telling something about herself and asking the apron recipient to correspond with her. Said Mrs. Craig, "If we can just describe to people the way we live, I'm sure that's all we will have to do to convince them that democracy is the most satisfactory form of government."[3]

Garments are represented extensively in all the instructional booklets intended to promote the use of cotton sacks. When Simplicity provided the designs, the garments featured were from the latest Simplicity line. Banning's research documents garments from cotton sacks from one woman's wardrobe and found that they were similar in design, construction, and fabric characteristics to fashions that appeared in *Good Housekeeping* magazine between the years 1949 and 1968.[4] While they might not have been considered the high fashion that would appear in such periodicals as *Vogue* or *Harper's Bazaar*, these patterns were fashionable particularly in rural areas or for women of middle and lower socioeconomic situations. The following images are dresses made of feed sack fabric.

Additionally, garments from feed sacks could be combined with a purchased garment or a garment made from purchased fabric. The images on pp. 78–80 show garments made from cotton feed sack fabric combined with garments of purchased fabric or purchased garments in the Museum's collection.

Skirts in the collection include those with applied waistbands and zippers, similar to any made from purchased fabric at the same time.

Accessories such as hair protectors / dust caps and make up or combing out capes were made for women from printed cotton sacks (examples follow).

Bonnets to protect women and girls from the sun while working outdoors were often made from feed sacks. This is logical since these garments were made for use at home and were likely not to be seen off the property.

▲ Rose and green floral dress of printed cotton sack fabric with buttons down the front, circa 1940. Gift of Lenna DeMarco, TTU-H2018-114-002.

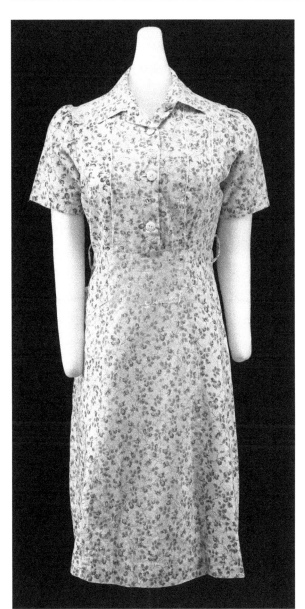

◄ Red flower print dress of printed cotton sack fabric with buttons down the front, circa 1940. Gift of Lenna DeMarco, TTU-H2018-114-004.

► Lilac and green floral dress of printed cotton sack fabric with buttons down the front and eyelet trim on collar, circa 1940. Gift of Lenna DeMarco, TTU-H2018-114-005.

◄ Blouse made from pieced feed sacks, circa 1940, TTU-H2015-053-001-022.

► Blouse made from feed sack printed with nautical images of sailboats, lighthouses, anchors, birds, and surf, circa 1940, TTU-H2015-053-001-034.

◄ A-line skirt made from feed sacks with a pink ground printed with a large floral pattern, circa 1940, TTU-H2015-053-001-014.

► This A-line skirt made of feed sacks printed with daisies closes with three bright yellow buttons, circa 1950, TTU-H2015-053-001-018.

◄ This skirt is made from panels of several different feed sack fabrics and has slit pockets, circa 1940, TTU-H2015-053-001-019.

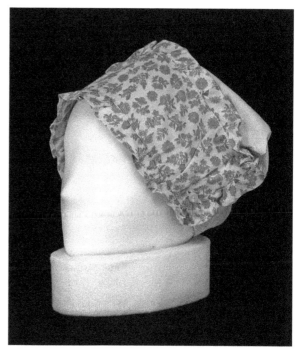

▲ Hair protector / dust cap of white feed sack and printed feed sack fabric, circa 1940. Gift of Lenna DeMarco, TTU-H2018-114-016.

► Make-up cape or combing cape of pink printed feed sack fabric that was to be used by a woman to protect her clothing while applying make-up or combing out her hair, circa 1940. Gift of Lenna DeMarco, TTU-H2018-114-008.

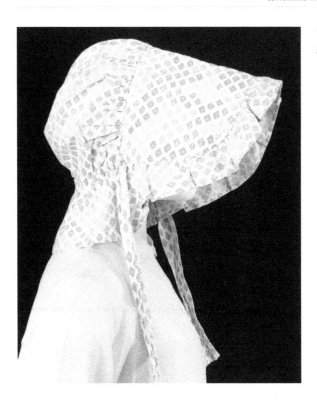

◄ Bonnet made of feed sack fabric printed with a geometric and floral pattern, circa 1940, TTU-H2015-053-001-040.

Children's clothing was ideal for using printed cotton sacks. Many children did not care about being in on the latest fashion, they outgrew their clothes quickly, and they were rough on their garments, so making children's clothing out of free printed cotton sack fabric was logical.

Sometimes children were allowed to go along to the feed store to select the bags that would be made into their clothing. Linda Timmons Fisher recalls going with her mother, Margaret Wilson Timmons, to the feed store on the Clovis Highway in Lubbock, Texas, to pick out fabrics. The family went to buy chicken feed for the chickens they kept at their home on Duke Street in Lubbock.[5] Examples of garments made from these sacks by Mrs. Timmons for her daughter, Linda, are shown on p. 84. Similar experiences of other children are recorded in publications on feed sacks. A particularly poignant memory from Joan Griffith of Smicksburg, Pennsylvania, was recorded by Adrosko:

I can remember the excitement I felt on Saturday mornings when I was allowed to go to the feed mill with my dad. While we were waiting for the grinding to be completed, I'd look at feed sacks and dream about a new dress.

When it came time for me to pick the sacks, I usually knew what I wanted. One of my favorite dresses was a pink flowered one that had a bustle in the back – Mom always kept up with fashions by looking at Sears catalogs.

When I was in sixth grade, I was voted the best dressed girl in my class. That was a real compliment to my mother because I was a poor country girl in a town school.[6]

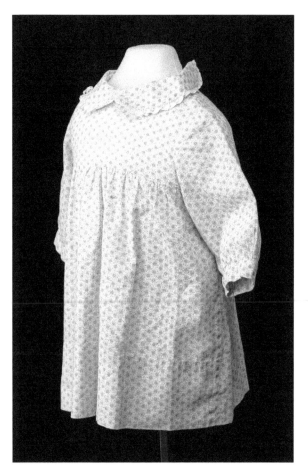

◄ Long-sleeved child's dress with lace trimmed collar made from feed sack fabric printed with a small floral pattern, circa 1940, TTU-H2015-053-001-023.
[06-028]

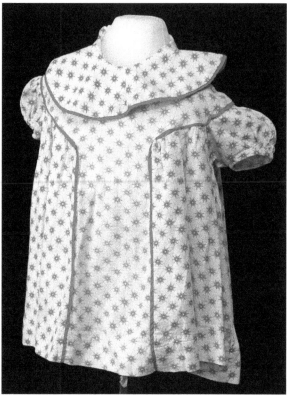

▲ This little girl's dress has puff sleeves and was made from feed sack fabric printed with red starbursts and trimmed with red bias binding, circa 1940, TTU-H2015-053-001-024.

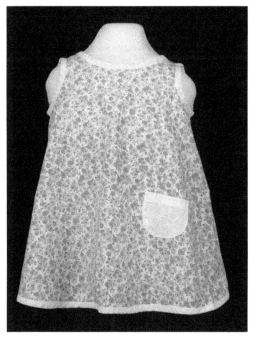

◄ This child's pinafore is made of feed sack fabric printed with a small floral pattern and has white bias binding at the edges and a white pocket, circa 1940, TTU-H2015-053-001-026.

This dress for a young girl was made from feed sack fabric printed with a floral pattern on a teal ground and is trimmed with lace down the front and on the pockets and ties in the back, circa 1940, TTU-H2015-053-001-032.

▲ Child's pink jumpsuit or pajamas of printed cotton sack fabric with buttons down the front, circa 1940. Gift of Lenna DeMarco, TTU-H2018-114-001.

◄ This child's shirt was made from feed sack fabric printed with a geometric pattern that was appropriate for either a little boy or girl, but, because of the way it buttons, was likely worn by a little girl, circa 1950, TTU-H2015-053-001-033.

◄ Child's printed feed sack apron trimmed with purchased binding made circa 1954 by Margaret Leatha Wilson Timmons (b. 1927, m. 1944, d. 1975; Mrs. T. L.) of Lubbock, Texas, for her daughter, Linda, born in 1949. Gift of Linda Timmons Fisher (Mrs. Charles) TTU-H2018-115-001.

► Green print feed sack crayon apron trimmed with purchased red binding made circa 1954 by Margaret Leatha Wilson Timmons (b. 1927, m. 1944, d. 1975; Mrs. T. L.) of Lubbock, Texas, for her daughter, Linda, born in 1949, is identical to the poncho shown in the *1954 Idea Book for Sewing with Cotton Bags*. The dates make it possible that Mrs. Timmons was inspired by this booklet. The donor recalls that her mother used feed sacks whenever she could and that she often accompanied her mother to pick out fabrics on the chicken feed sacks for her dresses. Gift of Linda Timmons Fisher (Mrs. Charles), TTU-H2018-115-002.

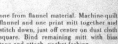

Decorative Place Mats

A friend who likes to entertain informally will appreciate these place mats. Make them from osnaburg bags, double-thickness, 8" by 12". Cut out large floral design from printed bag, and applique in one corner. Topstitch along fringed edges.

Dust Mitt

This handy dustcloth will brighten household chores. With pinking shears, cut square of cotton bag cloth 20" by 20". Make mitt pattern by drawing around hand. Cut 3 outlines, 2 from print and one from flannel material. Machine-quilt flannel and one print mitt together and stitch down, just off center on dust cloth square. Bind remaining mitt with bias tape and attach, pocket-fashion.

Rainy-Day Pair

Umbrella cover and matching bag are colorful accessories for rainy-day clothes. Measure umbrella, and cut cover, according to diagram, to fit snugly. Place wrong sides of material together, stitch seam, leaving 3" opening at top. Turn, stitch again to form French seam. Hem 3" opening at top. Hem lower edge.

the diagram to cut 2 circles of material for ruffle. Sew together, turn, and finish edges with top stitching or bias tape. Sew ruffle to umbrella cover. Turn hem down over ruffle edge and stitch. Fasten opening with hook and eye.

To make bag, cut piece of material 8½" by 18". Fold in half lengthwise. Stitch sides, leaving open about 3¼" from upper edge. Hem side edges left open. Turn upper edge down 1½" and stitch. Make another row of stitching about ½" above this to form casing for cords. Cut 2 cords 27" long. Insert cords in casing. Bag may be lined with any water-repellent fabric.

A new flour package— the 25-lb. pillowcase bag

Removable seam, just off center and down the full length of the pillowcase, divides container into two compartments. When the bag is empty, all you do is rip the seam and remove the label. Presto—you have a pillowcase! You can choose from plain white, white with pastel borders, or floral prints with solid borders.

Presenting the
cotton bag poncho

Cotton bags in 100-lb. sizes are made-to-order for ponchos, and you won't even need a pattern. We suggest that you wear these comfortable cover-ups around the house for working, playing, and entertaining.

To make a Cotton Bag Poncho:

Cut a 100-lb. bag into two lengths. Hem or bind edges. Sew shoulder seams at top. Add ties in the middle. Take your choice of necklines, straight, round, a V with lapels. Match a belt or sash to the lapels. Add your own decorative touches such as pockets, rickrack trimming, appliqued designs, or embroidery.

cotton bag aprons

This one-yard apron, in any one of four styles, can be made from a 100-lb. cotton bag. Simplicity 4443 (35¢) gives you a whole wardrobe of apron fashions, ranging from workaday to festive.

1. Both pretty and practical. Note scallops and trimming interest.

2. Sensible to work in. Wide pockets are well-placed and handy.

3. Strictly for show, with ruffles added to the scallops.

4. For the hostess. Becoming, in both living room and kitchen.

22

23

TTU-H2015-053-009-019

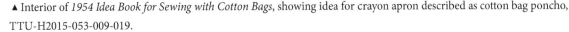

▲ Interior of *1954 Idea Book for Sewing with Cotton Bags*, showing idea for crayon apron described as cotton bag poncho, TTU-H2015-053-009-019.

◄ Young Linda Timmons, with her mother, Margaret Leatha Wilson Timmons (left) and grandmother, Margaret Dollie Wilson (right, holding Linda) circa 1951, slightly before the crayon apron shown above was made for her. Courtesy of Linda Timmons Fisher (Mrs. Charles).

Patterns for children's clothing were also printed on solid-colored pastel cotton sacks such as the one below, which shows a pattern for a pair of children's trousers. Doll or infant clothing is also represented in the collection.

A wide variety of garments, from full dresses to aprons, survive in the Museum's collection. There is no way to know if this is a true sampling of the garments that were made of feed sacks because, given the economic difficulties of the time and the population that generally used feed sack fabric for garments, garments were often worn until they were used up. A search through the instructional booklets in the Museum's collection did not reveal drawings of many garments made of feed sacks held in the collection. The 21 booklets dating from 1924 to 1970 comprise almost a full run of the booklets printed. In contrast, the garments that survive in the collection tend to be not as full and lack details that require more fabric and skill to make, such as ruffles, cuffs, and elaborate pocket styles. Therefore, many garments must have been made of feed sack fabrics using patterns other than those shown in the promotional instructional booklets.

▲ Too small for a child, this outfit was likely made for a doll from feed sack fabric printed in a floral and paisley pattern, circa 1940, TTU-H2015-053-001-036.

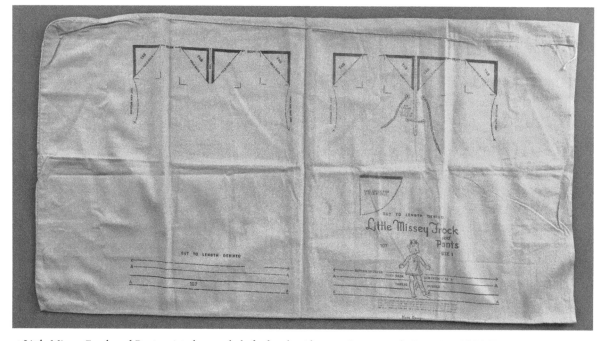

▲ Little Missey Frock and Pants printed on a whole feed sack with an apricot ground, circa 1940, TTU-H2015-053-003-697.

CHAPTER SEVEN ~~~~~ QUILTS FROM FEED SACKS

Quilt historians were among the first to begin the research on feed sacks because these fabrics appeared in many of the quilt documentations begun as part of the quilt revival around the U.S. Bicentennial in 1976. Several states initiated quilt documentations that resulted in significant books, which gave scholars the opportunity to consider regional differences in quilts. In many cases, images from these documentations have been uploaded to The Quilt Index, and more are to come. The Quilt Index at www.quiltindex.org is the only online, searchable index for a type of material culture object. In dating and grouping the quilts, not only did the beautifully printed fabrics from the cotton sacks show up, but also quilt backs and in some cases quilt fronts still bore the difficult-to-remove printing from white cotton sacks. The Museum's holdings include many quilts made from feed sack fabrics, including the plain white sacks that retained the advertising printing on them.

▲ Nine Patch quilt: the white fabric, which is probably a protective guard, is from flour sacks. The quilt was pieced by Abigail Dawkins Miller (Mrs. H.E.), the mother of the donor, 1890–1920, in Eric in the Oklahoma Territory. Gift of Myrtle Miller Austin (b. 1903, d. 1989; Mrs. Frank B.), TTU-H1978-139-034.

◄ Devil's Highway Quilt created by Mrs. John Vaughn of Matador, Texas, and finished by her sister after she died. The top was pieced circa 1890-1910 and quilted in 1946. The backing is made of home-dyed flour sack fabric from the Light Crust Flour Company. The backing was turned to the front to form the light blue binding. Gift of Mrs. G. C. Keith, TTU-H1979-180.

◄ Reverse of Devil's Highway Quilt showing logo from Light Crust Flour Company, which was not entirely removed. Gift of Mrs. G. C. Keith, TTU-H1979-180

Some quilt makers used dyed plain sacks or employed the pretty printed cotton sacks in beautifully pieced or appliqued blocks.

Ethel Abernathy was raised in Centre, Alabama in a farm family. On December 1, 1907 she married a neighborhood boy, Robert Abernathy, in Cherokee, Alabama. They had four daughters and a son. Robert was first a farmer and then a carpenter. They moved from Alabama to Floydada, Texas, in 1927, where the quilt below was made. In addition to being a prolific quilter, Ethel kept a beautiful flower and vegetable garden and was a great cook. Although partially made of "free" printed feed sack fabric, this quilt is an indication that the Abernathys were in a comfortable economic situation because Ethel was able to purchase all the background fabric. The elegantly curved edged border and fine quilting indicate that this quilt was not hastily sewn to keep the family warm but that the maker had the time to do beautiful work.

▲ Ethel Abernathy of Floydada, Texas. Photo courtesy of Judith Abernathy.

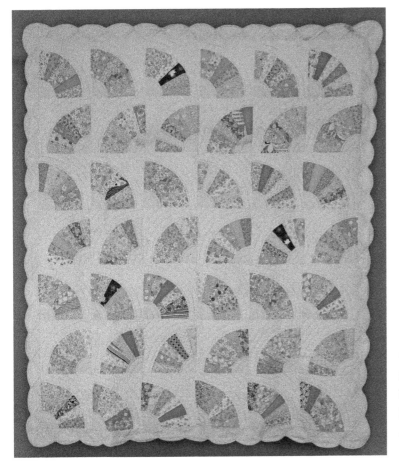

◄ Grandmother's Fan Quilt made by Melinda Ethel Abernathy ("Ethel," b. 1887, m. 1907, d. 1981; Mrs. Robert) in Floydada, Texas, circa 1940. Gift of Judith Abernathy, TTU-H2015-082-002.

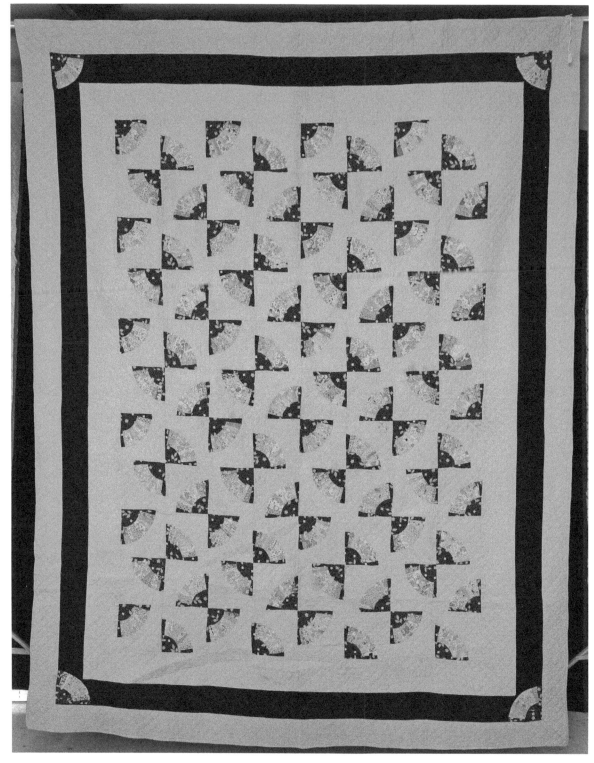

▲ This unique arrangement of the Grandmother's Fan pattern, employing black and yellow with an assortment of printed sack fabrics, is an indication of the artistic eye of the maker. The printed feed sack fabrics in the quilt date it to the late 1930s or early 1940s. Gift of Alice Larson, TTU-H2017-003-002.

◄ Sunbonnet Sue Quilt made by Margaret Dollie Wilson (b. 1912, m. 1928, d. 1963; Mrs. Harry) from printed feed sacks, which had first been made into dresses for her twin daughters, Margaret (b. 1927) and Marjorie (b. 1927). When the girls outgrew their dresses, their mother used the fabric to make this quilt, circa 1942. Gift of Linda Timmons Fisher (Mrs. Charles), TTU-H2015-084-006.

◄ Margaret Dollie Wilson with her twin daughters, Margaret Leatha Wilson, later Timmons (on left) and Marjorie Marie Wilson, later Martin (on right) who wore the dresses of printed cotton sacks that provided the fabric for the Sunbonnet Sue quilt when the girls outgrew them. Photo courtesy of Linda Timmons Fisher (Mrs. Charles).

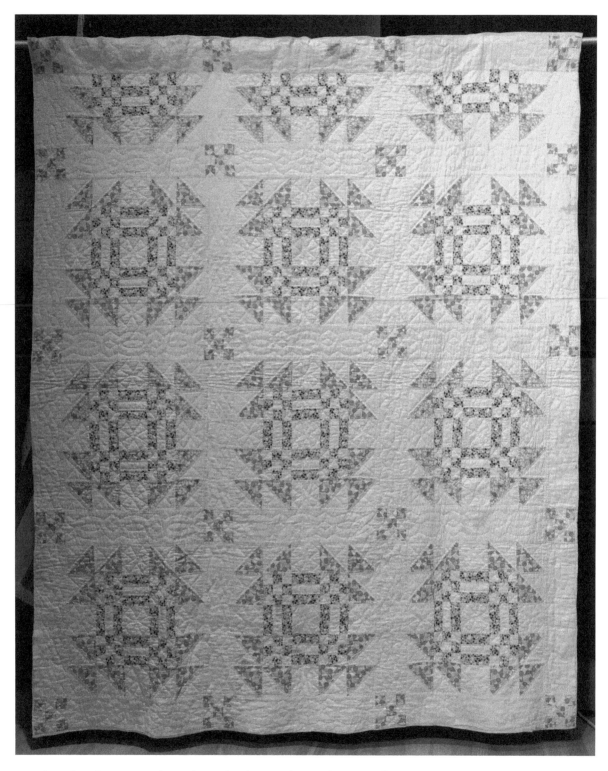

▲ Quilt made in the Goose in the Pond pieced quilt pattern by Cynthia Ann Biffle Locke Sweeney in Briscoe County, Texas, machine pieced, hand quilted. The feed sack fabric in the quilt and the maker's life dates the quilt circa 1940. Gift of Kathleen Hagaman Carson, TTU-H2018-003-001.

▲ Cynthia Ann Biffle Locke Sweeney (b. 1860, m. 1877, m. 1881, d. 1944) circa 1920 of Briscoe County, Texas, who made the Goose in the Pond quilt of printed cotton feed sacks. Photo courtesy of Kathleen Hagaman Carson.

The following quilts are from the Nickols Printed Cotton Sack Research Collection. Mrs. Nickols collected quilt tops and quilts for the fabrics and not necessarily because they were completed or beautiful quilts.

The Lone Star quilt with the feed sack back(see p. 95) from The William Kelly Milling Co. of Hutchinson, Kansas can be dated after 1905 when William Kelly established the company.[1] Although the exact date the Milling Company began their "Old Honesty" brand is not known, it was mentioned in the January 1914 issue of *The Southwestern Grain and Flour Journal.*[2]

Some quilts made from feed sack fabrics incorporated small pieces from a large number of printed cotton sacks and are considered scrap quilts. Other quilts made from feed sack fabrics incorporated small pieces from a large number of printed cotton sacks with one main fabric that came from one or more feed sacks such as the two examples shown on pp. 96–97.

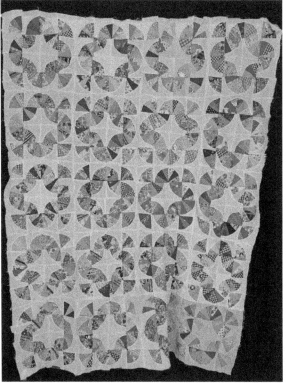

▲ The pattern in which style this quilt top is made is listed in Brackman's *Encyclopedia of Pieced Quilt Patterns* as #3369 with the names, Mohawk Trail, Path of Fans, Chinese Fan, and Baby Bunting. The printed fabrics are feed sack fabrics of floral, geometric, plaid, polka dot, and gingham patterns, circa 1950, TTU-H2015-053-002-033.

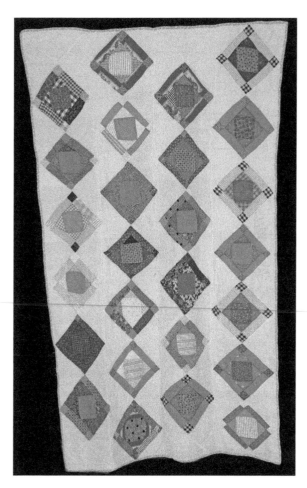

◄ This quilt uses feed sack fabrics printed in floral, plaid, stripe, and geometric patterns and has a floral pink printed binding. The block is a square on square similar to #2376 named Economy Patch, Hour Glass, Thrift Block, or This and That in Brackman's *Encyclopedia of Pieced Quilt Block Patterns*, circa 1940, TTU-H2015-053-002-043.

► This log cabin pieced quilt is filled with a wool blanket for batting. The blue sashing outlines the log cabin blocks pieced with floral, plaid, striped, and geometric printed feed sack fabric, circa 1940, TTU-H2015-053-002-044.

▲ This soft pastel example of a Lone Star quilt is backed with a flour sack from Old Honesty Flour whose advertising ink survives. Based on the pastel colors and the date for the beginning of production of Old Honesty Flour, the quilt was made circa 1925. TTU-H2015-053-002-045.

◀ Reverse detail showing the feed sack from Old Honesty Flour and The William Kelly Milling Company, used for the backing of the Lone Star quilt, TTU-H2015-053-002-045.

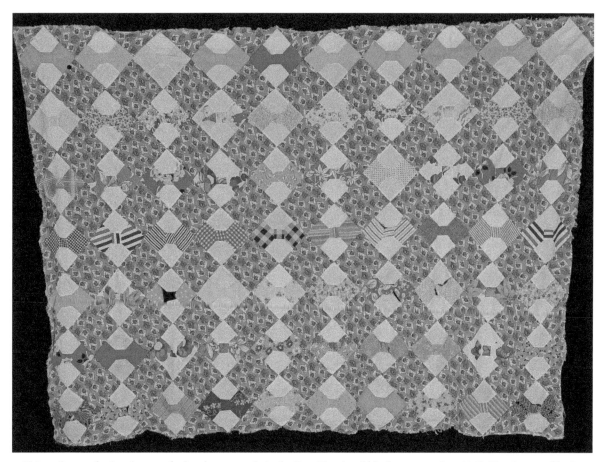

▲ This quilt top of printed cotton feed sacks was made in the Bow Tie pieced block pattern, circa 1940. The alternate blocks are all of the same printed cotton feed sack fabric in a floral rose pattern. The alternate blocks for a quilt this size might have been made from one feed sack, or it is possible that two matching sacks were needed. The feed sacks in this quilt top include floral, stripe, and geometric patterns, TTU-H2015-053-002-048.

▲ This quilt top made of feed sack fabrics uses a floral printed feed sack fabric for the alternate block, which would have required a significant amount of matching feed sacks. More than one feed sack was needed for the alternate blocks. This is a good example of the sort of quilt that required its maker to be sure she had enough fabric to complete a project. This quilt would seem to document the oral histories of quilters, and the men in their lives, who tell of searching for a specific fabric on a feed sack at the feed store. The Dresden plate pieced and appliqued blocks include floral, stripe, plaid, and geometric printed feed sack prints, circa 1940, TTU-H2015-053-002-049.

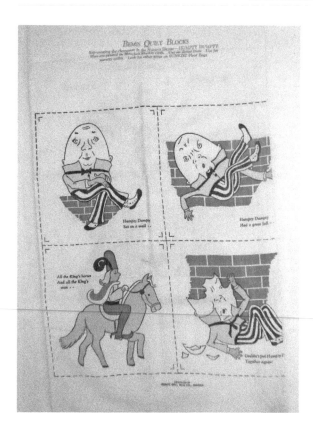

◄ Bemis Quilt Blocks representing the story of Humpty Dumpty on a flour sack for Sunkist Flour from the Maney Milling Company, Omaha, Nebraska, circa 1940, TTU-H2015-053-005-004.

◄ This bag was printed so that blocks representing the rural lifestyle could be cut from the back and used as quilt blocks. Additionally the sack is significant because the Red Star Milling Company, a subsidiary of General Mills, employed the popular dance band, The Texas Playboys, in a promotional scheme. Band member Johnnie Lee Wills negotiated a contract with the Red Star Milling Company to sponsor their radio show on KVOO in Tulsa, Oklahoma, in the fall of 1935. The wrappers for the flour were to contain pictures of various band members and their favorite recipes. This particular bag is likely an edition printed after the initial plan to show pictures of the band members with their recipes, circa1940–1950. Gift from the Good People of Austin, Texas, TTU-H2017-086-001.

◄ This partial feed sack, circa 1945, is printed with two quilt blocks likely intended for a child's alphabet book. The block on the top is printed "M is for mouse with 'Chester Cheese,'" and the lower block is printed "P is for pig with 'Paddy Porkchop.'" The blocks could be embroidered or appliqued or used in the book as is. Gift of Dottiemae and Harold Groves, TTU-H2018-018-002.

▲ ► Nine-patch quilt, an example of how one quilter used feed sacks on the back of a simple blue and white nine-patch quilt (see p. 101). Linda Fisher called her quilt, which she made in 2003, Sack Full of Nine Patches. This quilt is significant because it documents a wide variety of different manufacturers of products sold in twelve examples of cotton sacks. The manufacturers represented on the back of this quilt include (1) Pioneer Sugar, Michigan Sugar Company, Saginaw, Michigan; (2) Tropical Rose Cake Flour, Cherry Valley Mills, Cherry Valley, Arkansas; (3) Admiration, Oriental Style, Extra Fancy Long Grain Enriched Rice 103, Ricearama, A Supreme Oil Company, Product of U.S.A; (4) Willhite Seeds, Willhite Seed Company, Poolville, Texas; (5) Sunflower Seeds, Barlow Seed Company, G & J Barlow Proprietors Hastings, Michigan; (6) Westover Perfect Grass Lawn Seed, Westover Greenhouses; (7) Smokey Valley Best Flour, Smokey Valley Flour Mills, Lindsborg, Kansas; (8) Apple Seeds, Bowens Mills Apple Cider and Grist Mill, Middleville, Michigan; (9) Stock's Wheat Bran, F.W. Stock & Sons, Div. DCA Food Industries, Inc., Hillsdale, Michigan; (10) Admiration, Oriental Style 103 Extra Fancy Long Grain Enriched Rice, Ricearama, A Supreme Oil Company, Product of U.S.A; (11) Bendix Home Laundry; (12) Diamond Peerless Bleached Flour, Diamond Milling Co. Tyler, Texas. Gift of Linda Timmons Fisher (Mrs. Charles), TTU-H2016-015-005.

Some white sacks were printed with quilt blocks that could be cut out and made into quilts such as the three examples shown on pages 98 and 99.

Some quilt makers at the beginning of the twenty-first century continue to use feed sacks. An example from the Museum's collection of how one quilt maker used feed sacks bearing their original printed logos on the back of her quilt is shown on pages 100 and 101. A comparison of the hundreds of manufacturers represented in the Nickols collection found only one firm in both the Nickols collection and on the back of this quilt: Stock Wheat Bran, F.W. Stock & Sons. Hillsdale, Michigan. This sampling illustrates just how many different firms used cotton sacks to package their products.

◄ Linda Fisher (Mrs. Charles), who made the above Sack Full of Nine Patches quilt employing feed sacks for the back of the quilt. Photo courtesy of Linda Fisher.

CHAPTER EIGHT···THE WIDE VARIETY OF PRINTED FABRICS USED FOR COTTON SACKS

The first flour sacks made from dress-quality fabric were produced for Gingham Girl Flour in 1924.[1] Bags in pretty pastel tints were offered by the Staley Milling Company of Kansas City beginning in 1936 and were called "Tint-sax."[2] The bags were manufactured by Percy Kent Bag Company of a fine weave fabric and were offered in eleven different shades.[3]

Beginning in 1937, the Percy Kent Bag Company began putting prints on cotton sacks. Connolly, quoting from *PK. Our First Hundred Years*, an unpublished booklet of the Percy Kent Bag Co., Inc. in 1985, states,

> Richard K. Peek, Vice President of the Percy Kent Bag Company, is said to have 'discovered' the idea of using dress goods fabric for feed sacks while breakfasting in a Wichita, Kansas coffee shop in 1937. He noticed that the backs of the wooden chairs had cretonne slip covers in various pastel prints and thought, 'why wouldn't bags made out of these shades be a knockout with the housewife?'[4]

The date of the first printed feed sack fabrics is significant because it dates any garment or quilt made with printed cotton sack fabric as later than that initial printing date. The Percy Kent Bag Company places this crucial dating year at 1937. However, Rhodes states that she has identified forty-three references which document an earlier existence of feed sack prints, ranging between 1927 and 1937. Several of these dates are tied to memories of women regarding when they wore printed cotton sack garments as well as prints that survive in the Bemis

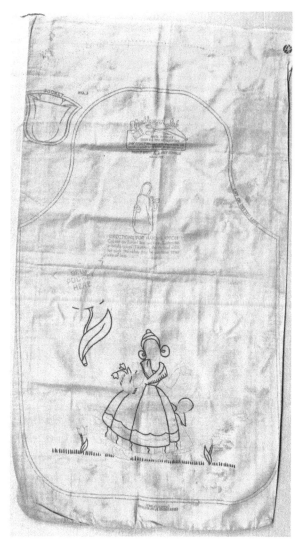

▲ Teal feed sack with printed label Globe Tint-sax, copyright 1938, printed with directions for making an apron, produced by Percy Kent Art Service, TTU-H2015-053-003-310.

Bag Manufacturing Company archives.[5] This level of documentation would argue for feed sack prints being available certainly by 1930. As discussed earlier, this was not the first time sacks were made from dress-quality fabric—the Gingham Girl and the Tint-sax were from dress-quality fabric—but this was the first time that beautiful prints were put on the fabric for making sacks. Printed feed sack fabric is not as tightly woven as fabrics available for purchase in the twenty-first century for quilting and clothing. Given the advances in the textile industry, this is not unexpected. There is a slight coarse quality to many of the sack fabrics. Fabrics from the sacks can definitively be identified by the large holes left from the stitching along the bag edges. However, Banning comments that once the manufacturers began packing products in bags of printed fabric, women found the fabric similar to fabric that could be purchased off the bolt because some of the same textile mills who sold fabric to cotton sack manufacturers also sold the same fabric on bolts to retail stores.[6]

The textile bag manufacturing industry quickly recognized the popularity of the sacks the Percy Kent Bag Company was producing, and all of them moved into design and production of their own line of fabrics for cotton sacks. This decision significantly improved the companies' economic situations and once again moved cotton sacks ahead of paper bags because of consumer demand. Another significant factor in the increased use of cotton was the mechanical cotton picker, which was introduced in 1936. Prior to that time, all cotton was picked by hand.[7] Cook reports that consumption of cotton fabrics in bags during this shift climbed from 816 million yards in 1939 to 1,283 million yards in 1946.[8] Clearly the appeal of the beautiful prints was a factor in this increased consumption. Although the beautiful fabric added five to seven cents per bag to the feed store-owner's purchase costs, the bags were in such demand that feed store owners believed that they had to stock them.[9] Relevant information obtained from oral histories of feed storeowners appears in Appendix D of Fawn Valentine's *West Virginia Quilts and Quiltmakers; Echoes From the Hills*. Seldom cited in the literature on feed sacks, these oral histories provide firsthand accounts of the types of fabrics used in the sacks and the popularity of the sacks.

A variety of fibers, weaves, finishes, and fabrics were used for the bags. As mentioned, the fabric used for the bags came from the same textile manufacturers who also provided fabrics to the apparel industry and yardage to be cut from the bolt at a dry goods store. James Cannon, a North Carolina textile manufacturer, remembered that there was a certain pattern of domestic cloth that most of his female customers wanted. He called it "cannon cloth," which was light enough in weight for women's clothing and heavy enough for sacking. Sheeting in various weights was made up into fabric for greige-goods bags and print-good bags for many years. Although cotton was the chief fiber source for these fabrics, other fibers were used, such as flax, burlap, rayon, nylon, and kenaf (a jutelike fiber). Fine cotton prints and rayon sheeting were especially popular for bags that were later used for many household items. With this diversity considered, looking at these textile bags as if only just one type of fabric is to miss the wide variety that were manufactured.[10]

When heavier, sturdier bags were required for packaging the sack's contents, the fabric was woven of yarns that were highly twisted in opposite directions. The resulting fabrics were generally not desirable for clothing or quilt tops, but were sometimes used to back utility quilts.[11] The most commonly used fabric for cotton feed sacks was "print cloth," or a balanced-weave cotton known as muslin. Muslin came in different weights and yarn counts. High thread count print cloth was called percale and was always used for flour sacks and later for feed sacks. Home sewers and quilt makers favored this fabric. When cambric was used for flour sacks, it had a highly sized finish to prevent dusting of the flour through the fabric. When washed, cambric was very soft and was often used for baby clothes. Denim was

also used for bags with the intention that it could be made into clothes for men and boys.[12]

Various weaves were employed for feed sack fabrics, including basket and twill. Some fabrics were even woven with a metallic yarn and were highly prized for high school evening gowns. Bags that came later in the phenomenon included those of fashionable fabric such as dimity,[13] pique,[14] broadcloth[15] and seersucker.[16]

Mr. Foster, owner of the Foster Feed store in Weston, West Virginia, reported that in the mid-1930s he bought print bags from Eishelman Red Rose Company in two sizes, one for horse and mule feed and another for chicken mash. These cost around 40 cents each. The bag was filled at the mill and closed with a hand-held electric sewing machine. The thread used to close it was cotton until WWII when it changed to nylon. Some instructional booklets, such as *A Bag of Tricks for Home Sewing* published in 1942–1943 included information on how to crochet the cotton thread from the cotton bags to make a doily or a bedspread.[17]

Some mill owners did not pay the additional cost to package their goods in pretty printed sacks.[18] However, Wade Beckman, feed storeowner and manager of Beckman Brothers Incorporated in West Virginia was one who saw the value of paying the additional cost. Beckman said that he prided himself on carrying the best product and what the customer wanted: they wanted printed bags. Beckman said the cost was $1.00 to $1.50 extra per ton of feed, averaging 5-7 cents per bag in the early 1940s. Additionally, Beckman described an exchange system in which his customers could return their bags for a 25-cent deposit, which he resold at the same price. Mr. Beckman believed it was good business to provide this service without a profit: those who lived in town and did not have access to the feed sacks used on the farms often came in to purchase the empty bags. On average he recalled that there were usually 25-100 returned bags per week from which town residents could choose.[19] The instructional booklets also pointed consumers to this type of source for printed cotton sacks with a section similar to the following:

How and where to find cotton bags:

If cotton bags are new to you, you wonder where you can find them. If you are a cotton bag fan of long standing, you may be hunting new sources of supply. The chief outlets for cotton bags are:

1. Feed stores in farming communities. Buy your feed and the bag comes free. A few dealers sell empty bags.
2. Bakeries use flour in 100-lb. bags. These bags may be plain or print. Some bakeries sell them across the counter, or on house-to-house bread routes. Ask your baker if he has bags for sale.
3. Groceries and super markets sell plain and print bags. It pays to inquire if you do not see a display of bags for sale. Family sizes of meal, salt, and flour are sold in cotton bags.
4. Department stores sell white cotton bags for tea towels.
5. Bag converters buy empty bags from bakers and advertise them for mail-order resale.
6. Farmer's markets.
7. Classified section of telephone directory.

If there isn't a single place to buy cotton bags in your community, and you can't find them in your mail-order catalog, write to the Textile Bag Manufacturers Association, 611 Davis Street, Evanston, Illinois. They will give you a list of firms from whom you may order cotton bags.[20]

Oral interviews provide insight into what was available in the area of Morgantown, West Virginia, and dispel the idea that overnight all cotton sacks came with pretty prints. Glen Core recalled that all food products came in paper packaging, but the feed came in both printed and plain white branded bags. Hugh Phillips sold salt, sugar, and flour in cot-ton bags but didn't remember having printed bags. He did recall that the bags had premiums of towels, dishes, and even toys in them. Walter Raese, who operated groceries in both Davis and Morgantown, recalled that beans came in 100-pound burlap bags, while sugar was in a burlap bag with a cotton cloth lining. According to Raese, salt was in cloth and paper because it would draw dampness and get wet. Flour came in both white branded and printed cotton bags and was put in bins in the store and sold to the consumer in paper bags. But some families purchased the 100-pound bag rather than the smaller quantity. Raese remembers both farmers and moonshiners bought 100-pound bags of sugar.[21]

Charlotte Williams remembers going with her mother to purchase feed sacks for the fabrics in Spur, Texas where she grew up on a dry land farm.

I went with my daddy to the Star Feed Store in Spur which was owned by M. H. Bragg, to get the chicken feed so I could pick out the printed sacks I liked. It took two sacks to make me a gathered skirt, which I made myself. I probably did this from about the age of 12 until I graduated high school in 1958 and moved away. My mother made her house dresses, aprons, pillow cases, cup towels, quilt tops and curtains from feed sacks, as we could not afford to purchase fabric and there was very little selection in Spur. Mother also made dresses for my little sister, although not as many, as the weather changed and farming improved and they could afford more 'store bought' clothing for the three younger kids.[22]

On p. 107 are some of the most interesting printed pattern examples from Charlotte Williams' feed sack fabric donation representing the fabrics her family used for clothing, quilts and household items.

Textile bag manufacturers sought to increase the desirability of printed cotton sacks by looking to New York designers to design the fabrics in the 1940s. The

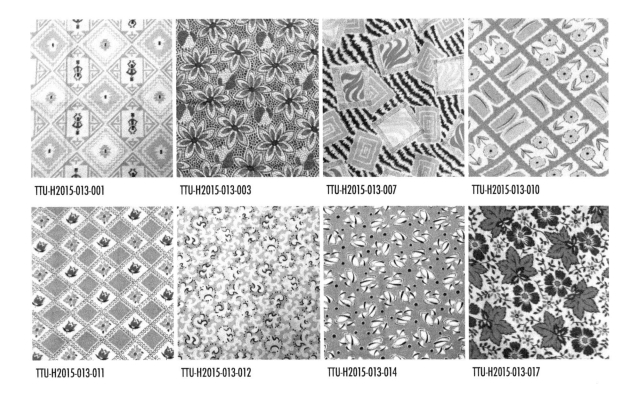

| TTU-H2015-013-001 | TTU-H2015-013-003 | TTU-H2015-013-007 | TTU-H2015-013-010 |
| TTU-H2015-013-011 | TTU-H2015-013-012 | TTU-H2015-013-014 | TTU-H2015-013-017 |

Percy Kent Bag Company hired A. Charles Barton as their design director whose studio was in the fashion hub New York City. A Percy Kent ad referenced Barton's involvement: "From his studio in New York, Mr. Barton, one of America's foremost designers, sends out the distinctive ideas for which Percy Kent Bags are famous." The ad further described Mr. Barton as "European by birth and education", who had "won wide recognition . . . as one of America's foremost fabric designers."[23] Barton toured the American Midwest, to view first-hand the many uses to which Ken-Print [the Percy Kent printed cotton sack fabric] was put to use and to get new ideas for future fabric prints. The idea being to make Ken-Print bags the 'glamour sacks' of America.[24] Erwin was another firm designing fabrics to be placed on cotton sacks.

The prints employed for cotton sacks range from polka dots and stripes to multi-colored prints. The multi-colored designs included florals, geometrics, abstracts, stripes, plaids, novelty, juvenile, western, toile, and propaganda prints. Border prints were used for tablecloths, but also on occasion for skirts. Sacks were also made from fabric with a printed quilt block pattern, often called cheater cloth or pre-printed quilt patterns. Newcome and Nessleroad say that the timesaving cheater cloth product came about as women took on the dual roles of homemaker and wage earner.[25] Coordinating fabrics of plain and print weave were also produced, some that seem specifically intended for curtains with a ruffle at the top. The more than 3300 swatches plus the 381 partial and 300 complete printed feed sacks in the Museum of Texas Tech University's collection provide a comprehensive array of the types of fabrics available to the consumer.

One of the more unique prints made by the Percy Kent firm was a propaganda piece showing the point of view of the Allies during World War II, known as the Battle for Britain (see p. 110). Although other significant events of World War II are printed on the

◄ Women in Erwin dress print dresses from ad circa 1940.

▼ Cotton sack with Erwin Dress print design from Gulf-Way Feeds. Erwin Dress Prints is documented in the upper right-hand corner of the paper label, circa 1940, TTU-H2015-053-003-253.

▲ Example of what is called cheater cloth or preprinted pieced quilt block in a pinwheel pattern on the Polar Bear Self-Rising Flour sack, circa 1955, TTU-H2015-053-002-053.

fabric, this is generically called the Battle for Britain. The fabric commemorates Allied forces and has negative images for the Axis powers.

World War II influenced the production of cotton sacks with the War Production Board (WPB) forcing conversion of about 88 percent of the cotton textile industry to war or essential civilian production. Cotton needed to take the place of burlap, which had been imported from India, to fill the need for bagging for sand bags, camouflage cloth, and food and agricultural bags. The WPB ordered from 20 to 100 percent of the weaving looms in textile mills shifted to the war effort, which would account for slightly more than 88 per cent of total capacity.[26] In 1942 the Texas Commissioner of Agriculture cautioned farmers to save their feed or seed bags. Because of the war's curtailment of shipping, there was a scar-

city of bagging material and farmers were encouraged to help the defense program by conserving the available supply of feed sacks and sending them back into circulation by returning them to their feed dealer, feed manufacturer, or to any bag dealer.[27] However, there is evidence that there were still bags to be had for making garments and home furnishings: the National Cotton Council continued to put out booklets about how to use the bags, 4-H clubs were still teaching girls how to make garments from the sacks,[28] and Home Demonstration Clubs held fashion shows with up-to-the minute styles made from feed sacks.[29] This was likely because textile rationing during World War II did not originally apply to feed sacks, which were classified as "industrial" textile products."[30]

Feed sacks played a role in the plan designed by researchers to establish adequate standards of clothing for a family of five with an income of $1,500.00.[31] With the food rationing effort necessitated by World War II, even urban dwellers turned to raising their own victory gardens and chickens,[32] so these households would have had printed cotton fabric from the chicken feed cotton sacks. After World War II, relief efforts around the world included gifts of feed sacks along with cotton clothing and sewing kits sent to Manila,[33] and the sacks were still in demand for clothing in the United States.[34] A 1949 article in the *Atlanta Daily World* reported that a young Tennessee woman showed the "Sunday" dress she made from the feed sacks she acquired while caring for her 400 chickens.[35]

In 1953 The Fulton Bag and Cotton Mills in Atlanta, Georgia developed cotton bags designed for reuse as pillowcases, luncheon cloths, toweling, and other useful items. The pillowcase style came on the market in August of 1953 and by October the Fulton Company couldn't keep up with demand.

Fulton used "an ingenious false stitch, turning out only full-sized pillowcases, regardless of whether the bag contained 25 or 50 pounds of a product."[36] Although headquartered in Atlanta, the firm had

▲ Battle for Britain propaganda fabric designed by the Percy Kent design firm, circa 1944, TTU-H2015-053-005-010a.

◄ Detail of Battle for Britain fabric by the Percy Kent Bag Company showing Hitler, Mussolini and Hirohito, leaders of the countries opposing the Allies, in a frying pan with the caption, "Keep 'em frying!" The image also lionizes the leadership of General Douglas Mac Arthur, shows the Russian Bear stopping the Nazi advance at Stalingrad, and commemorates the bombing of Pearl Harbor, TTU-H2015-053-005-010a.

▲ Cotton sack printed to be used as a pillowcase with figures representing people from the Netherlands and tulips, circa 1955, TTU-H2015-053-003-592.

▲ King Midas printed this flour sack to be used as a pillowcase after the flour was emptied, circa 1955, TTU-H2015-053-003-587.

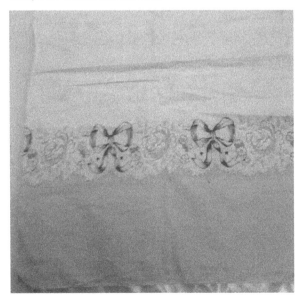

▲ Feed sack printed to be sewn into a pillowcase, possibly made by the Fulton Bag and Cotton Mills in Atlanta, Georgia, circa 1955. Gift of Virginia Hernandez, TTU-H2016-023-001.

plants in nine cities, including Dallas, in the heart of the Texas cotton-growing region. The plant was built in 1906 and located near Fair Park,[37] where the State Fair of Texas regularly showcased Texas cotton products. The plant processed an estimated railcar of cotton cloth every other day into bags of all sizes and designs. The plant did its own printing and engraving on the cotton fabric as well as on paper for the labels. In addition to the tightly woven fabric for pillowcases, they turned out the porous bags used for citrus and onion crops, burlap bags, furniture bags, cotton sacks, tarpaulins for tents and football field covers, and dozens of other items. [38]

Similar to the Fulton Bag Company's idea of printing sack fabric to be sewn into pillowcases was the gift given to two newlyweds in Lubbock, Texas. In 1953, when Nancy married her husband Robert Fehleison, she was given many brand new feed sacks bearing the label "Stanton's Stock & Poultry Feeds; Manufactured by The Standard Milling Co., Lubbock, TX" as a wedding gift from the firm. Her husband was the son of the owners of The Standard Milling Company, and Nancy was expected to make curtains for their new home from the fabric. The story is illustrative of the difference between the expectations of brides in the 1950s and those of today. Nancy Fehleison didn't mind, and from the feed sack fabric made café curtains with ruffles at the top for their kitchen. Mrs. Fehleison donated three of the original bags that she did not use, in two different prints, to the Museum.

Other home accessory items made of feed sacks included clothespin bag and casserole covers.

▲ Stanton feed sack from The Standard Milling Company, Lubbock, Texas, 1953, given to Nancy Fehleison as a wedding gift from her new husband's family business so that she could make curtains for their new home. Gift of Nancy Fehleison TTU-H2015-012-002.

► Robert and Nancy Fehleison on their wedding day, Sunday afternoon, June 21, 1953. Photo courtesy of the children of Nancy and Robert Fehleison.

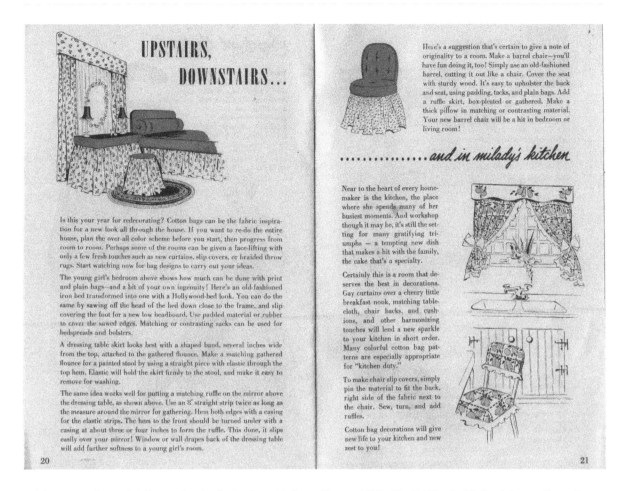

UPSTAIRS, DOWNSTAIRS...

Is this your year for redecorating? Cotton bags can be the fabric inspiration for a new look all through the house. If you want to re-do the entire house, plan the over-all color scheme before you start, then progress from room to room. Perhaps some of the rooms can be given a face-lifting with only a few fresh touches such as new curtains, slip covers, or braided throw rugs. Start watching now for bag designs to carry out your ideas.

The young girl's bedroom above shows how much can be done with print and plain bags—and a bit of your own ingenuity! Here's an old-fashioned iron bed transformed into one with a Hollywood-bed look. You can do the same by sawing off the head of the bed down close to the frame, and slip covering the foot for a new low headboard. Use padded material or rubber to cover the sawed edges. Matching or contrasting sacks can be used for bedspreads and bolsters.

A dressing table skirt looks best with a shaped band, several inches wide from the top, attached to the gathered flounce. Make a matching gathered flounce for a painted stool by using a straight piece with elastic through the top hem. Elastic will hold the skirt firmly to the stool, and make it easy to remove for washing.

The same idea works well for putting a matching ruffle on the mirror above the dressing table, as shown above. Use an 8″ straight strip twice as long as the measure around the mirror for gathering. Hem both edges with a casing for the elastic strips. The hem to the front should be turned under with a casing at about three or four inches to form the ruffle. This done, it slips easily over your mirror! Window or wall drapes back of the dressing table will add further softness to a young girl's room.

20

Here's a suggestion that's certain to give a note of originality to a room. Make a barrel chair—you'll have fun doing it, too! Simply use an old-fashioned barrel, cutting it out like a chair. Cover the seat with sturdy wood. It's easy to upholster the back and seat, using padding, tacks, and plain bags. Add a ruffle skirt, box-pleated or gathered. Make a thick pillow in matching or contrasting material. Your new barrel chair will be a hit in bedroom or living room!

...............and in milady's kitchen

Near to the heart of every homemaker is the kitchen, the place where she spends many of her busiest moments. And workshop though it may be, it's still the setting for many gratifying triumphs — a tempting new dish that makes a hit with the family, the cake that's a specialty.

Certainly this is a room that deserves the best in decorations. Gay curtains over a cheery little breakfast nook, matching tablecloth, chair backs, and cushions, and other harmonizing touches will lend a new sparkle to your kitchen in short order. Many colorful cotton bag patterns are especially appropriate for "kitchen duty."

To make chair slip covers, simply pin the material to fit the back, right side of the fabric next to the chair. Sew, turn, and add ruffles.

Cotton bag decorations will give new life to your kitchen and new zest to you!

21

▲ These pages from *1952 Pattern Service for Sewing with Cotton Bags*, published in 1952 by the National Cotton Council, document the types of curtains Nancy Fehleison was expected to make from the cotton sacks she received as a wedding gift, TTU-H2015-053-017.

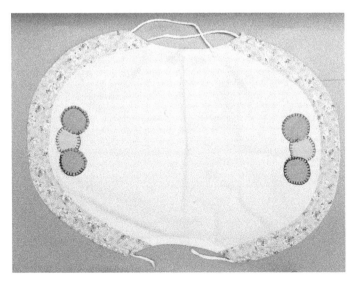

◄ Casserole cover of white feed sack fabric, edged with casing of printed feed sack fabric with drawstrings, circa 1940. Gift of Lenna DeMarco, TTU-H2018-114-017.

▲ Clothespin bag designed to look like a child's dress of white fabric and printed red and grey feed sack fabric, circa 1940. Gift of Lenna DeMarco, TTU-H2018-114-006.

▼ Clothespin bag of red white and blue printed feed sack fabric, trimmed with purchased red bias binding and red rickrack, circa 1940. Gift of Lenna DeMarco, TTU-H2018-114-007.

The printed fabric swatches that make up the remainder of this catalogue are drawn from the more than 4,100 different fabric swatches in the Nickols collection. This number does not include the examples of prints on the collection on the printed cotton feed sacks which are part of the 942 full and partial cotton sacks in the Nickols collection. These printed cotton fabric examples were printed possibly as early as 1927 but more likely between 1937 and 1970. These swatches are organized according to the subject of the print and include abstracts or geometric, animals, florals, foods, novelty and stripes or plaids.

The Decline of the Cotton Feed Sack

Cook discusses the pricing differences that fueled the rise of the paper bag manufacturers that would come to dominate the market in the late 1940s, ultimately ceding victory to paper bags. The paper industry promoted their bags, constructed of four walls of heavy Kraft paper, as the best in price and sanitation. By 1948 cotton bags began to be priced out of the market by paper bags., and paper bags accounted for over 53% of the shipping-sack market. In that same year, 1948, the price of 100 three-multiwall paper bags cost little more than the price of one single cotton bag. Before World War II, raw cotton cost 10 cents per pound, and in 1948 the price had risen to about 36 cents per pound. Meanwhile unbleached wood pulp had increased from 3 ¼ cents per pound to only about 6 cents per pound. Cotton bags cost 32 cents for a 100-pound bag verses 10 cents for the same size paper bag.[39]

Leon Droll, who was in upper management at the bag manufacturer Bemis, outlined another reason why the cotton sacks went out of fashion: "After the 1960s cotton bags were phased out mainly because families were smaller and women were working so they didn't cook as much and didn't buy large bags of flour."[40] The development of synthetic fibers also contributed to the decline of cotton bag production.[41]

Although printed cotton sacks are not in production today for consumer goods, twenty-first century quilters or seamstresses looking for reproduction fabric to make quilts and garments with a feed sack look have reproduction sack fabric available. The plain white cotton sacks that were reused through ingenuity and often made beautiful with embroidery and other embellishments, along with the popular pretty printed sacks, certainly brightened the lives of those who worked with, used, and wore or slept under the sack fabric. While some recall wearing clothing made of feed sacks as a sign of poverty, many others recall those days with a fondness and pride in their family's careful husbandry of resources. The cotton feed sack, then, can certainly be valued today as an American artifact embodying the principles of thrift, industry, and ingenuity.

FEED SACK PRINTS – ABSTRACT AND GEOMETRICS

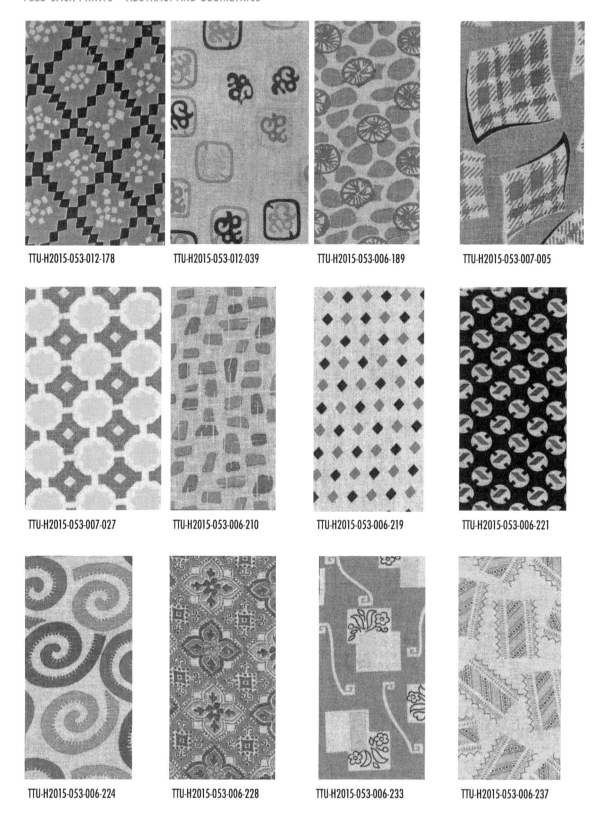

TTU-H2015-053-012-178   TTU-H2015-053-012-039   TTU-H2015-053-006-189   TTU-H2015-053-007-005

TTU-H2015-053-007-027   TTU-H2015-053-006-210   TTU-H2015-053-006-219   TTU-H2015-053-006-221

TTU-H2015-053-006-224   TTU-H2015-053-006-228   TTU-H2015-053-006-233   TTU-H2015-053-006-237

FEED SACK PRINTS – ABSTRACT AND GEOMETRICS

TTU-H2015-053-006-239

TTU-H2015-053-006-241

TTU-H2015-053-006-243

TTU-H2015-053-006-245

TTU-H2015-053-006-246

TTU-H2015-053-006-251

TTU-H2015-053-006-257

TTU-H2015-053-006-273

TTU-H2015-053-003-049

TTU-H2015-053-003-061

TTU-H2015-053-003-072

TTU-H2015-053-003-092

FEED SACK PRINTS – ABSTRACT AND GEOMETRICS

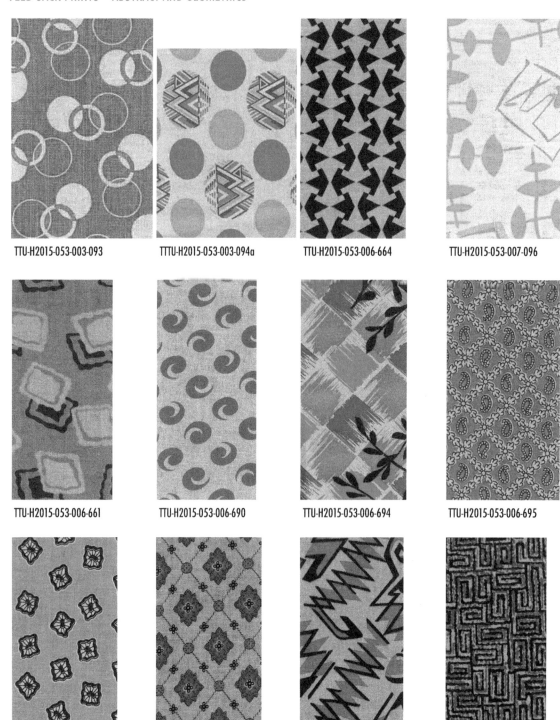

TTU-H2015-053-003-093   TTTU-H2015-053-003-094a   TTU-H2015-053-006-664   TTU-H2015-053-007-096

TTU-H2015-053-006-661   TTU-H2015-053-006-690   TTU-H2015-053-006-694   TTU-H2015-053-006-695

TTU-H2015-053-006-685   TTU-H2015-053-006-683   TTU-H2015-053-006-677   TTU-H2015-053-006-788

FEED SACK PRINTS – ABSTRACT AND GEOMETRICS

TTU-H2015-053-006-779

TTU-H2015-053-006-770

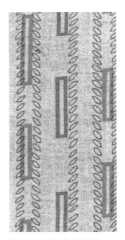
TTU-H2015-053-006-761

TTU-H2015-053-006-750

TTU-H2015-053-006-728

TTU-H2015-053-006-719

TTU-H2015-053-006-336

TTU-H2015-053-006-855

TTU-H2015-053-006-699

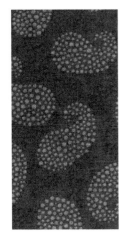
TTU-H2015-053-006-784

TTU-H2015-053-007-017

TTU-H2015-053-006-350

FEED SACK PRINTS – ABSTRACT AND GEOMETRICS

TTU-H2015-053-006-277	TTU-H2015-053-006-272	TTU-H2015-053-007-251	TTU-H2015-053-007-182
TTU-H2015-053-007-180	TTU-H2015-053-007-175	TTU-H2015-053-006-500	TTU-H2015-053-006-368
TTU-H2015-053-006-359	TTU-H2015-053-006-322	TTU-H2015-053-006-365	TTU-H2015-053-006-362

FEED SACK PRINTS – ABSTRACT AND GEOMETRICS

TTU-H2015-053-006-407

TTU-H2015-053-006-282

TTU-H2015-053-006-414

TTU-H2015-053-006-288

TTU-H2015-053-006-308

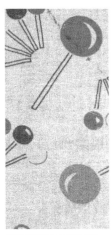

TTU-H2015-053-006-313

TTU-H2015-053-006-366

TTU-H2015-053-006-326

TTU-H2015-053-006-335

TTU-H2015-053-006-704

TTU-H2015-053-006-278

TTU-H2015-053-006-396

FEED SACK PRINTS – ABSTRACT AND GEOMETRICS

TTU-H2015-053-006-403

TTU-H2015-053-006-413

TTU-H2015-053-006-415

TTU-H2015-053-006-418

TTU-H2015-053-006-419

TTU-H2015-053-003-530b

TTU-H2015-053-003-816

TTU-H2015-053-003-819

TTU-H2015-053-006-456

TTU-H2015-053-006-457

TTU-H2015-053-006-458

TTU-H2015-053-006-474

FEED SACK PRINTS – ABSTRACT AND GEOMETRICS

TTU-H2015-053-006-558

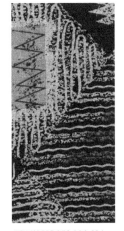

TTU-H2015-053-006-494

TTU-H2015-053-006-501

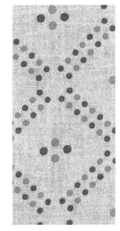

TTU-H2015-053-006-503

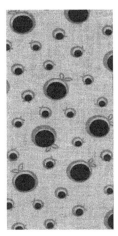

TTU-H2015-053-006-504

TTU-H2015-053-006-506

TTU-H2015-053-006-523

TTU-H2015-053-006-561

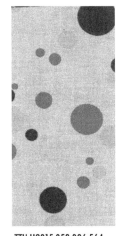

TTU-H2015-053-006-564

TTU-H2015-053-006-566

TTU-H2015-053-006-572

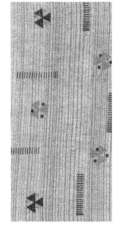

TTU-H2015-053-006-575

FEED SACK PRINTS – ABSTRACT AND GEOMETRICS

TTU-H2015-053-006-579

TTU-H2015-053-006-587

TTU-H2015-053-006-609

TTU-H2015-053-006-617

TTU-H2015-053-006-656

TTU-H2015-053-006-657

TTU-H2015-053-006-659

TTU-H2015-053-006-693

TTU-H2015-053-006-686

TTU-H2015-053-003-505

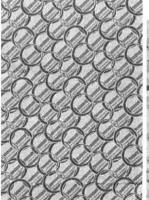

TTU-H2015-053-003-666

TTU-H2015-053-003-813

FEED SACK PRINTS – ABSTRACT AND GEOMETRICS

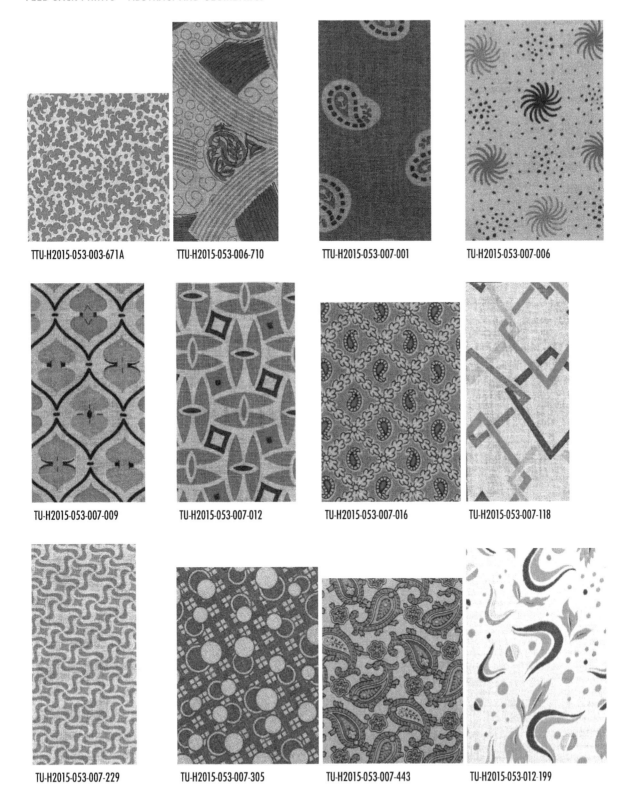

TTU-H2015-053-003-671A   TTU-H2015-053-006-710   TTU-H2015-053-007-001   TU-H2015-053-007-006

TU-H2015-053-007-009   TU-H2015-053-007-012   TU-H2015-053-007-016   TU-H2015-053-007-118

TU-H2015-053-007-229   TU-H2015-053-007-305   TU-H2015-053-007-443   TU-H2015-053-012-199

FEED SACK PRINTS – ABSTRACT AND GEOMETRICS

TU-H2015-053-012-204

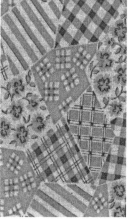

TU-H2015-053-007-956

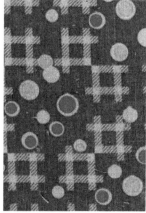

TU-H2015-053-007-989

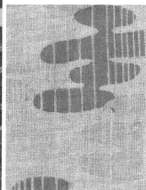

TU-H2015-053-007-997

TTU-H2015-053-008-597

TTU-H2015-053-008-725

TTU-H2015-053-012-205

TTU-H2015-053-012-206

FEED SACK PRINTS – ANIMALS, BIRDS, BUTTERFLIES, ETC.

TTU-H2015-053-012-121

TTU-H2015-053-012-120

TTU-H2015-053-012-118

TTU-H2015-053-003-229

TTU-H2015-053-003-212

TTU-H2015-053-006-899

TTU-H2015-053-006-738

TTU-H2015-053-006-713

TTU-H2015-053-006-238

TTU-H2015-053-006-467

TTU-H2015-053-008-128

TTU-H2015-053-006-702

FEED SACK PRINTS – ANIMALS, BIRDS, BUTTERFLIES, ETC.

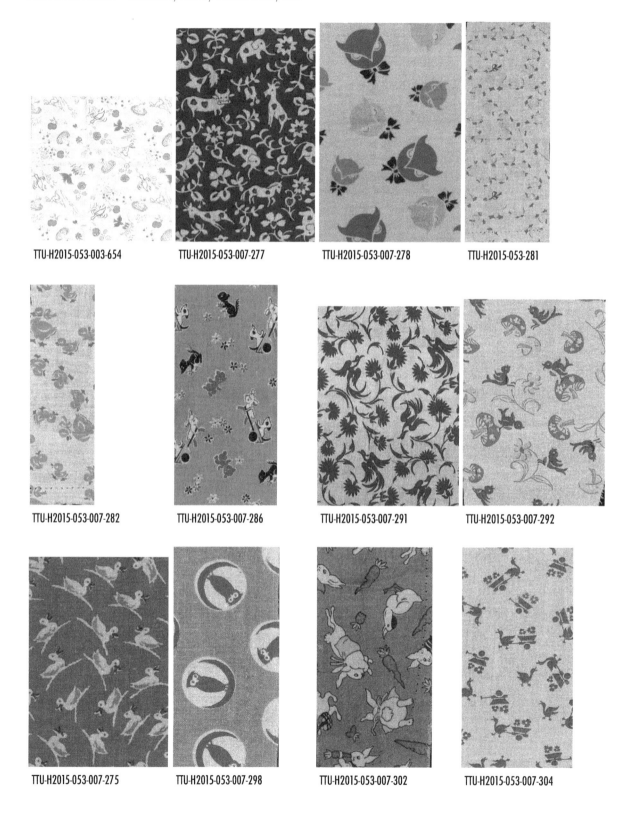

TTU-H2015-053-003-654

TTU-H2015-053-007-277

TTU-H2015-053-007-278

TTU-H2015-053-281

TTU-H2015-053-007-282

TTU-H2015-053-007-286

TTU-H2015-053-007-291

TTU-H2015-053-007-292

TTU-H2015-053-007-275

TTU-H2015-053-007-298

TTU-H2015-053-007-302

TTU-H2015-053-007-304

FEED SACK PRINTS – ANIMALS, BIRDS, BUTTERFLIES, ETC.

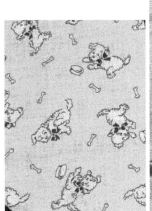

TTU-H2015-053-007-386

TTU-H2015-053-012-106

TTU-H2015-053-007-394

TTU-H2015-053-007-405

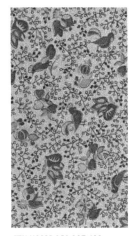

TTU-H2015-053-007-490

TTU-H2015-053-007-493

TTU-H2015-053-012-119

TTU-H2015-053-012-095

TTU-H2015-053-007-839

TTU-H2015-053-007-843

TTU-H2015-053-012-116

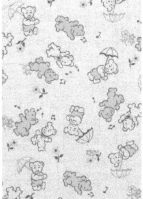

TTU-H2015-053-008-565

FEED SACK PRINTS – ANIMALS, BIRDS, BUTTERFLIES, ETC.

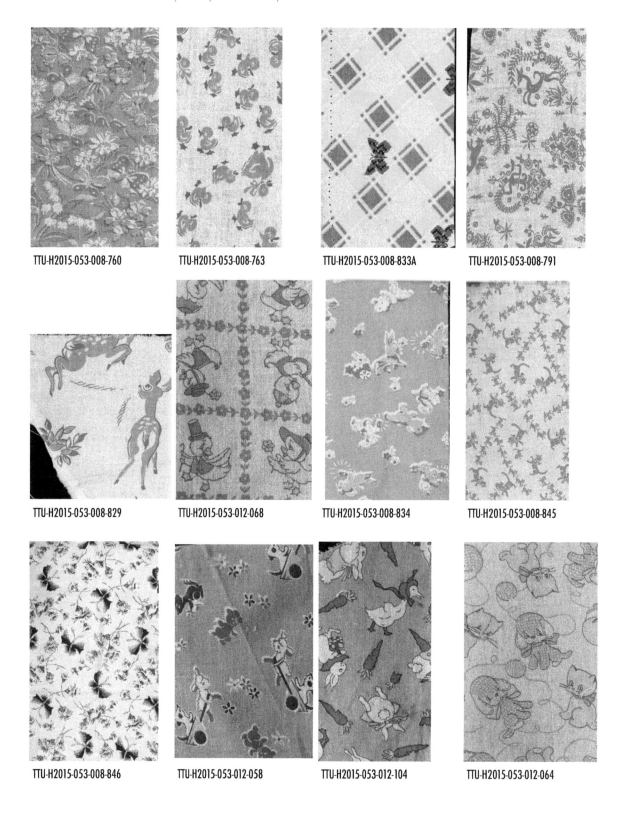

TTU-H2015-053-008-760

TTU-H2015-053-008-763

TTU-H2015-053-008-833A

TTU-H2015-053-008-791

TTU-H2015-053-008-829

TTU-H2015-053-012-068

TTU-H2015-053-008-834

TTU-H2015-053-008-845

TTU-H2015-053-008-846

TTU-H2015-053-012-058

TTU-H2015-053-012-104

TTU-H2015-053-012-064

FEED SACK PRINTS – ANIMALS, BIRDS, BUTTERFLIES, ETC.

TTU-H2015-053-012-096

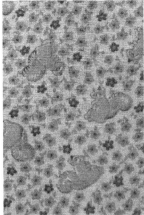
TTU-H2015-053-012-100

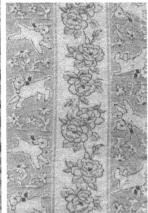
TTU-H2015-053-012-102

TTU-H2015-053-012-105

TTU-H2015-053-012-107

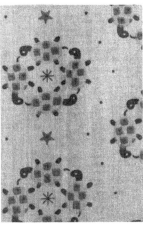
TTU-H2015-053-012-108

TTU-H2015-053-012-109

TTU-H2015-053-012-112

TTU-H2015-053-012-113

TTU-H2015-053-012-114

TTU-H2015-053-012-115

TTU-H2015-053-007-852

FEED SACK PRINTS – ANIMALS, BIRDS, BUTTERFLIES, ETC.

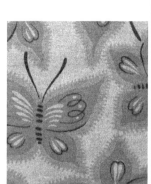

TTU-H2015-053-012-117

TTU-H2015-053-012-123

TTU-H2015-053-012-126

TTU-H2015-053-012-127

TTU-H2015-053-012-128

TTU-H2015-053-012-130

TTU-H2015-053-008-584

FEED SACK PRINTS – FLORAL

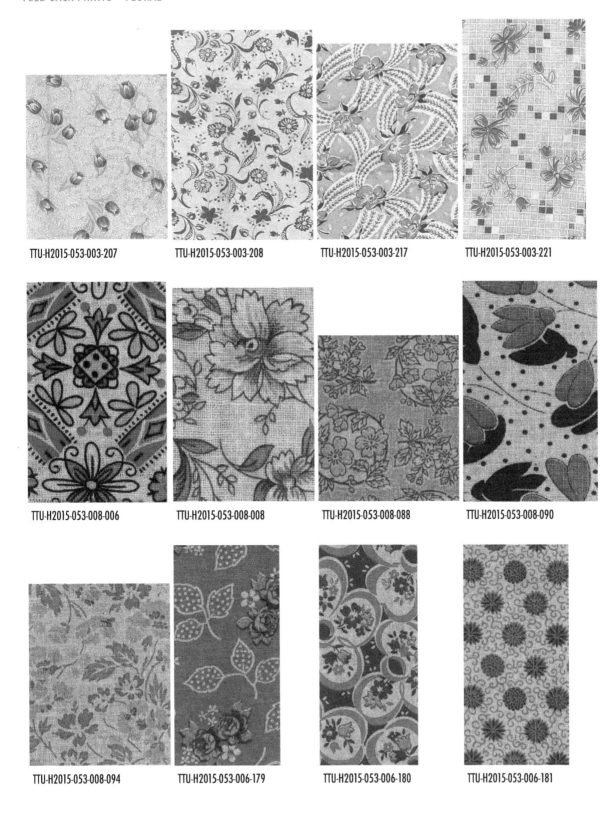

TTU-H2015-053-003-207

TTU-H2015-053-003-208

TTU-H2015-053-003-217

TTU-H2015-053-003-221

TTU-H2015-053-008-006

TTU-H2015-053-008-008

TTU-H2015-053-008-088

TTU-H2015-053-008-090

TTU-H2015-053-008-094

TTU-H2015-053-006-179

TTU-H2015-053-006-180

TTU-H2015-053-006-181

FEED SACK PRINTS – FLORALS

TTU-H2015-053-006-182

TTU-H2015-053-006-183

TTU-H2015-053-006-185

TTU-H2015-053-006-186

TTU-H2015-053-012-037

TTU-H2015-053-012-038

TTU-H2015-053-006-190

TTU-H2015-053-007-002

TTU-H2015-053-006-194

TTU-H2015-053-006-198

TTU-H2015-053-006-199

TTU-H2015-053-006-203

FEED SACK PRINTS – FLORALS

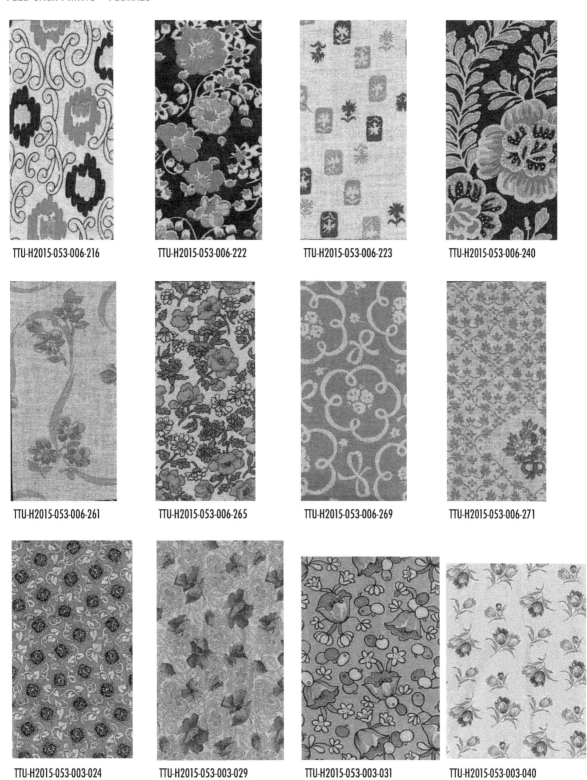

TTU-H2015-053-006-216

TTU-H2015-053-006-222

TTU-H2015-053-006-223

TTU-H2015-053-006-240

TTU-H2015-053-006-261

TTU-H2015-053-006-265

TTU-H2015-053-006-269

TTU-H2015-053-006-271

TTU-H2015-053-003-024

TTU-H2015-053-003-029

TTU-H2015-053-003-031

TTU-H2015-053-003-040

FEED SACK PRINTS – FLORALS

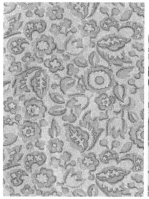

TTU-H2015-053-003-041

TTU-H2015-053-003-050

TTU-H2015-053-003-053b

TTU-H2015-053-003-054

TTU-H2015-053-003-055

TTU-H2015-053-003-063

TTU-H2015-053-003-066b

TTU-H2015-053-003-076

TTU-H2015-053-003-077

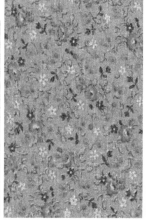

TTU-H2015-053-003-079

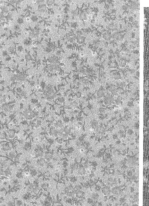

TTU-H2015-053-003-081

TTU-H2015-053-003-105

FEED SACK PRINTS – FLORALS

TTU-H2015-053-003-106

TTU-H2015-053-003-202

TTU-H2015-053-006-651

TTU-H2015-053-007-088

TTU-H2015-053-006-680

TTU-H2015-053-006-763

TTU-H2015-053-006-757

TTU-H2015-053-006-756

TTU-H2015-053-006-731

TTU-H2015-053-006-716

TTU-H2015-053-006-848

TTU-H2015-053-007-025

FEED SACK PRINTS – FLORALS

TTU-H2015-053-006-312

TTU-H2015-053-006-295

TTU-H2015-053-006-298

TTU-H2015-053-006-301

TTU-H2015-053-006-305

TTU-H2015-053-006-294

TTU-H2015-053-006-316

TTU-H2015-053-006-321

TTU-H2015-053-006-323

TTU-H2015-053-006-337

TTU-H2015-053-006-338

TTU-H2015-053-006-341

FEED SACK PRINTS – FLORALS

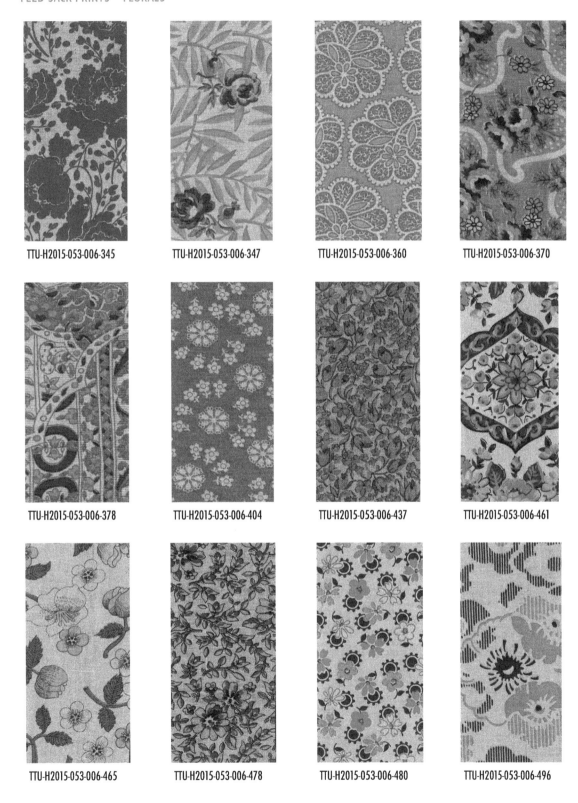

TTU-H2015-053-006-345

TTU-H2015-053-006-347

TTU-H2015-053-006-360

TTU-H2015-053-006-370

TTU-H2015-053-006-378

TTU-H2015-053-006-404

TTU-H2015-053-006-437

TTU-H2015-053-006-461

TTU-H2015-053-006-465

TTU-H2015-053-006-478

TTU-H2015-053-006-480

TTU-H2015-053-006-496

FEED SACK PRINTS – FLORALS

TTU-H2015-053-006-509

TTU-H2015-053-006-520

TTU-H2015-053-006-525

TTU-H2015-053-006-527

TTU-H2015-053-006-534

TTU-H2015-053-006-536

TTU-H2015-053-006-546

TTU-H2015-053-006-638

TTU-H2015-053-006-602

TTU-H2015-053-006-605

TTU-H2015-053-006-608

TTU-H2015-053-006-613

FEED SACK PRINTS - FLORALS

TTU-H2015-053-006-619

TTU-H2015-053-006-637

TTU-H2015-053-006-581

TTU-H2015-053-006-642

TTU-H2015-053-006-646

TTU-H2015-053-006-653

TTU-H2015-053-006-611

TTU-H2015-053-006-642

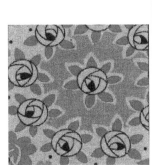

TTU-H2015-053-008-145

TTU-H2015-053-008-147

TTU-H2015-053-008-784

TTU-H2015-053-008-864

FEED SACK PRINTS – FOOD

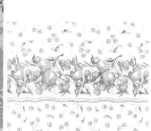

TTU-H2015-053-003-034

TTU-H2015-053-003-012

TTU-H2015-053-003-070

TTU-H2015-053-003-590

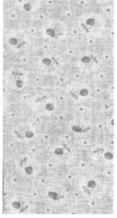

TTU-H2015-053-004-055

TTU-H2015-053-007-115

TTU-H2015-053-006-247

TTU-H2015-053-007-069

TTU-H2015-053-006-978

TTU-H2015-053-006-801

TTU-H2015-053-006-688

TTU-H2015-053-006-671

FEED SACK PRINTS – FOOD

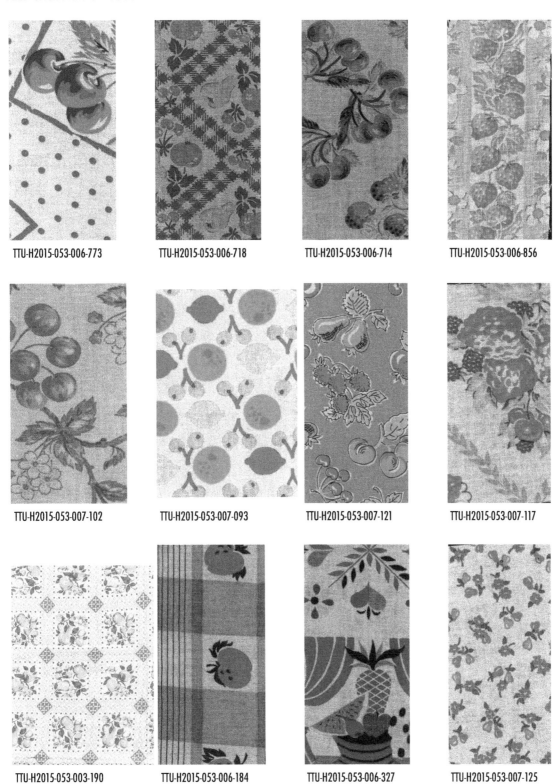

TTU-H2015-053-006-773

TTU-H2015-053-006-718

TTU-H2015-053-006-714

TTU-H2015-053-006-856

TTU-H2015-053-007-102

TTU-H2015-053-007-093

TTU-H2015-053-007-121

TTU-H2015-053-007-117

TTU-H2015-053-003-190

TTU-H2015-053-006-184

TTU-H2015-053-006-327

TTU-H2015-053-007-125

FEED SACK PRINTS – FOOD

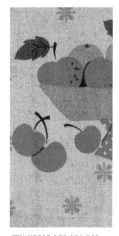

TTU-H2015-053-006-331

TTU-H2015-053-006-426

TTU-H2015-053-006-200

TTU-H2015-053-007-187

TTU-H2015-053-006-441

TTU-H2015-053-006-473

TTU-H2015-053-006-551

TTU-H2015-053-006-592

TTU-H2015-053-006-606

TTU-H2015-053-006-607

TTU-H2015-053-006-623

TTU-H2015-053-006-817

FEED SACK PRINTS - FOOD

TTU-H2015-053-003-665A

TTU-H2015-053-007-195

TTU-H2015-053-007-196

TTU-H2015-053-007-406

TTU-H2015-053-007-409

TTU-H2015-053-007-412

TTU-H2015-053-007-414

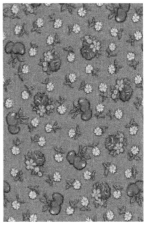

TTU-H2015-053-007-422

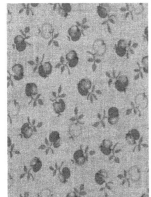

TTU-H2015-053-007-454

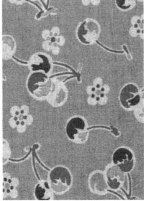

TTU-H2015-053-007-528

TTU-H2015-053-007-686

TTU-H2015-053-008-141

FEED SACK PRINTS – FOOD

TTU-H2015-053-008-146

TTU-H2015-053-008-152A

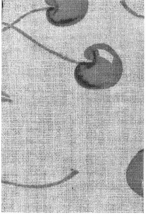

TTU-H2015-053-008-158

TTU-H2015-053-008-556

TTU-H2015-053-008-570

TTU-H2015-053-012-214

FEED SACK PRINTS – NOVELTY

TU-H2015-053-003-211a

TTU-H2015-053-003-213b

TTU-H2015-053-002-228

TTU-H2015-053-003-238

TTU-H2015-053-006-193

TTU-H2015-053-003-035

TTU-H2015-053-003-039

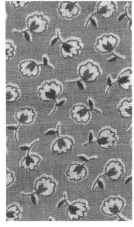

TTU-H2015-053-008-800

TTU-H2015-053-003-067

TTU-H2015-053-003-216

TTU-H2015-053-003-216?

TTU-H2015-053-003-082

FEED SACK PRINTS - NOVELTY

TTU-H2015-053-007-279

TTU-H2015-053-003-089c

TTU-H2015-053-003-107

TTU-H2015-053-003-108b

TTU-H2015-053-006-987

TTU-H2015-053-006-665

TTU-H2015-053-006-895

TTU-H2015-053-006-675

TTU-H2015-053-006-218

TTU-H2015-053-006-790

TTU-H2015-053-007-201

TTU-H2015-053-006-205

FEED SACK PRINTS – NOVELTY

TTU-H2015-053-006-696

TTU-H2015-053-006-229

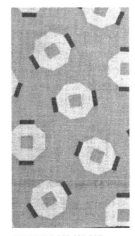

TTU-H2015-053-012-235

TTU-H2015-053-006-354

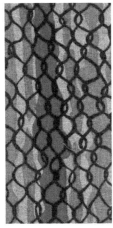

TTU-H2015-053-006-376

TTU-H2015-053-006-392

TTU-H2015-053-006-393

TTU-H2015-053-006-420

TTU-H2015-053-006-433

TTU-H2015-053-006-495

TTU-H2015-053-006-502

TTU-H2015-053-006-533

FEED SACK PRINTS – NOVELTY

TTU-H2015-053-006-612

TTU-H2015-053-006-674

TTU-H2015-053-003-585

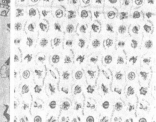

TTU-H2015-053-003-584

TTU-H2015-053-012-203

TTU-H2015-053-003-588

TTU-H2015-053-003-598

TTU-H2015-053-003-599

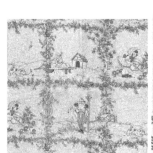

TTU-H2015-053-003-603

TTU-H2015-053-003-605

TTU-H2015-053-003-620

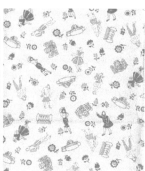

TTU-H2015-053-003-633

FEED SACK PRINTS – NOVELTY

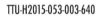

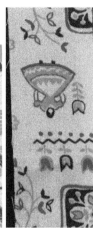

TTU-H2015-053-003-640   TTU-H2015-053-003-641   TTU-H2015-053-003-673   TTU-H2-15-053-008-103

TTU-H2015-053-007-107A   TTU-H2015-053-007-256G   TTU-H2015-053-007-268   TTU-H2015-053-007-270

TTU-H2015-053-007-274   TTU-H2015-053-007-276   TTU-H2015-053-007-280   TTU-H2015-053-007-283

FEED SACK PRINTS – NOVELTY

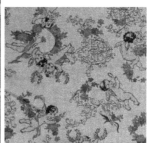

TTU-H2015-053-007-284

TTU-H2015-053-007-285

TTU-H2015-053-007-287

TTU-H2015-053-007-288

TTU-H2015-053-007-289

TTU-H2015-053-007-293

TTU-H2015-053-007-294

TTU-H2015-053-007-295

TTU-H2015-053-007-303

TTU-H2015-053-007-378

TTU-H2015-053-012-071

TTU-H2015-053-007-389

FEED SACK PRINTS – NOVELTY

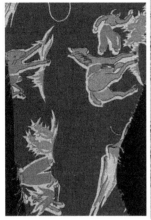

TTU-H2015-053-007-392

TTU-H2015-053-007-393

TTU-H2015-053-007-404

TTU-H2015-053-007-407

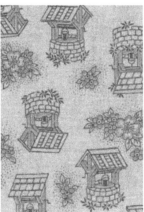

TTU-H2015-053-007-410

TTU-H2015-053-007-413

TTU-H2015-053-007-415

TTU-H2015-053-007-418

TTU-H2015-053-007-419

TTU-H2015-053-007-421

TTU-H2015-053-007-427

TTU-H2015-053-007-484

FEED SACK PRINTS – NOVELTY

TTU-H2015-053-007-559

TTU-H2015-053-012-061

TTU-H2015-053-007-652

TTU-H2015-053-007-653

TTU-H2015-053-007-666

TTU-H2015-053-012-093

TTU-H2015-053-007-844

TTU-H2015-053-007-864A

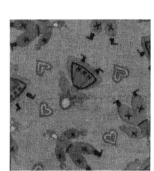

TTU-H2015-053-008-130A

TTU-H2015-053-008-135B

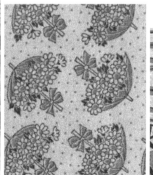

TTU-H2015-053-008-537

TTU-H2015-053-008-544

FEED SACK PRINTS – NOVELTY

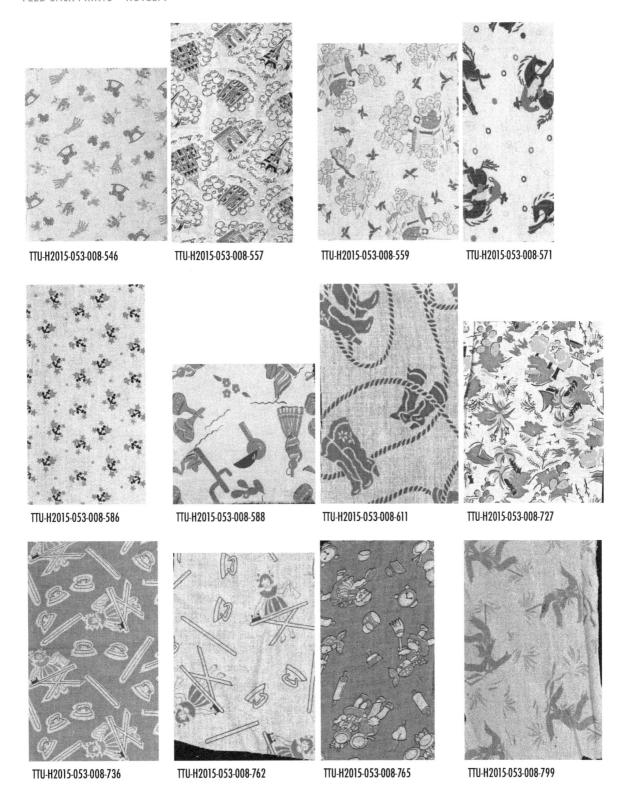

TTU-H2015-053-008-546

TTU-H2015-053-008-557

TTU-H2015-053-008-559

TTU-H2015-053-008-571

TTU-H2015-053-008-586

TTU-H2015-053-008-588

TTU-H2015-053-008-611

TTU-H2015-053-008-727

TTU-H2015-053-008-736

TTU-H2015-053-008-762

TTU-H2015-053-008-765

TTU-H2015-053-008-799

FEED SACK PRINTS – NOVELTY

TTU-H2015-053-008-811

TTU-H2015-053-008-812

TTU-H2015-053-008-816

TTU-H2015-053-008-831

TTU-H2015-053-008-832

TTU-H2015-053-008-836

TTU-H2015-053-012-037

TTU-H2015-053-012-038

TTU-H2015-053-012-042

TTU-H2015-053-012-043

TTU-H2015-053-012-053

TTU-H2015-053-012-054

FEED SACK PRINTS – NOVELTY

TTU-H2015-053-012-055

TTU-H2015-053-012-062

TTU-H2015-053-012-065

TTU-H2015-053-012-067

TTU-H2015-053-012-069

TTU-H2015-053-012-070

TTU-H2015-053-012-072

TTU-H2015-053-012-073

TTU-H2015-053-012-074

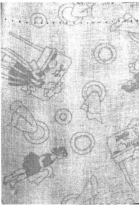

TTU-H2015-053-012-080

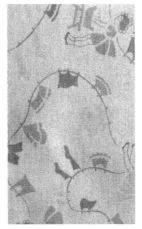

TTU-H2015-053-012-081

TTU-H2015-053-012-084

FEED SACK PRINTS – NOVELTY

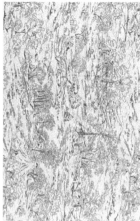

TTU-H2015-053-012-092   TTU-H2015-053-012-097   TTU-H2015-053-003-586

FEED SACK PRINTS – STRIPES AND PLAIDS

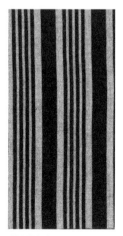

TTU-H2015-053-006-196   TTU-H2015-053-006-212   TTU-H2015-053-006-213   TTU-H2015-053-006-252

FEED SACK PRINTS – STRIPES AND PLAIDS

TTU-H2015-053-007-411

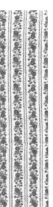

TTU-H2015-053-003-002a

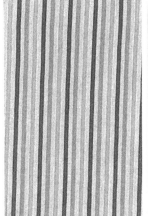

TTU-H2015-053-003-005

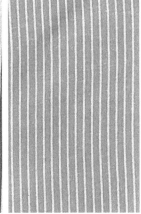

TTU-H2015-053-003-006

TTU-H2015-053-003-013

TTU-H2015-053-003-038b

TTU-H2015-053-003-064

TTU-H2015-053-003-087

TTU-H2015-053-006-684

TTU-H2015-053-006-682

TTU-H2015-053-006-762

TTU-H2015-053-006-755

FEED SACK PRINTS – STRIPES AND PLAIDS

TTU-H2015-053-006-739

TTU-H2015-053-006-737

TTU-H2015-053-006-736

TTU-H2015-053-006-735

TTU-H2015-053-006-734

TTU-H2015-053-006-729

TTU-H2015-053-006-720

TTU-H2015-053-006-864

TTU-H2015-053-006-700

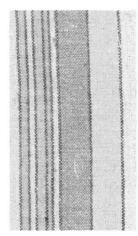

TTU-H2015-053-007-098

TTU-H2015-053-007-089

TTU-H2015-053-007-086

FEED SACK PRINTS – STRIPES AND PLAIDS

TTU-H2015-053-006-306

TTU-H2015-053-006-823

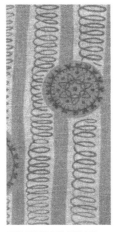

TTU-H2015-053-006-888

TTU-H2015-053-006-797

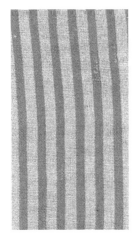

TTU-H2015-053-006-984

TTU-H2015-053-007-675

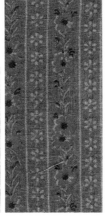

TTU-H2015-053-006-431

TTU-H2015-053-006-258

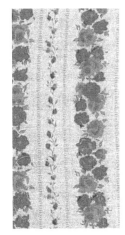

TTU-H2015-053-006-319

TTU-H2015-053-006-257

TTU-H2015-053-007-992

TTU-H2015-053-006-352

FEED SACK PRINTS – STRIPES AND PLAIDS

TTU-H2015-053-006-290

TTU-H2015-053-006-361

TTU-H2015-053-006-320

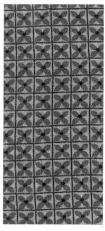
TTU-H2015-053-006-654

TTU-H2015-053-006-353

TTU-H2015-053-006-293

TTU-H2015-053-006-377

TTU-H2015-053-006-382

TTU-H2015-053-006-386

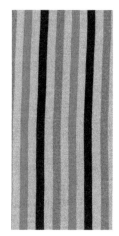
TTU-H2015-053-006-394

TTU-H2015-053-007-258D

TTU-H2015-053-006-828

FEED SACK PRINTS – STRIPES AND PLAIDS

TTU-H2015-053-006-460	TTU-H2015-053-006-462	TTU-H2015-053-006-470	TTU-H2015-053-006-472
TTU-H2015-053-006-477	TTU-H2015-053-006-554	TTU-H2015-053-006-555	TTU-H2015-053-006-556
TTU-H2015-053-006-557	TTU-H2015-053-006-559	TTU-H2015-053-006-618	TTU-H2015-053-006-632

FEED SACK PRINTS – STRIPES AND PLAIDS

TTU-H2015-053-006-578

TTU-H2015-053-006-624

TTU-H2015-053-006-626

TTU-H2015-053-006-639

TTU-H2015-053-006-640

TTU-H2015-053-003-674

TTU-H2015-053-003-634

TTU-H2015-053-003-812

TTU-H2015-053-003-514

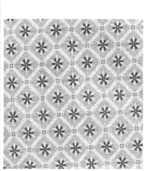

TTU-H2015-053-003-816

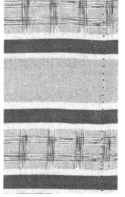

TTU-H2015-053-008-823

TTU-H2015-053-006-699

FEED SACK PRINTS – STRIPES AND PLAIDS

TTU-H2015-053-006-726

TTU-H2015-053-007-010

TTU-H2015-013-007

TTU-H2015-053-007-024

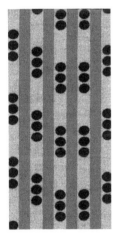

TTU-H2015-053-006-197

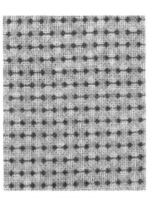

TTU-H2015-053-007-048

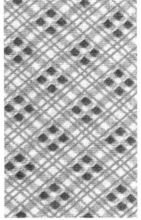

TTU-H2015-053-007-074

TTU-H2015-053-007-091

TTU-H2015-053-007-094

TTU-H2015-053-007-112

TTU-H2015-053-007-113

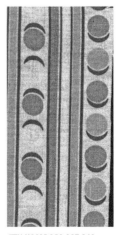

TTU-H2015-053-007-140

FEED SACK PRINTS – STRIPES AND PLAIDS

TTU-H2015-053-007-203

TTU-H2015-053-007-237

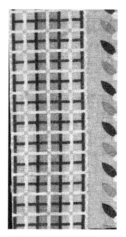

TTU-H2015-053-007-252

TTU-H2015-053-007-296A

TTU-H2015-053-007-473

TTU-H2015-053-007-509

TTU-H2015-053-007-529

TTU-H2015-053-007-641

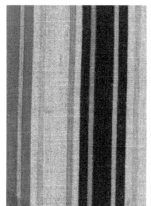

TTU-H2015-053-007-642

TTU-H2015-053-007-797

TTU-H2015-053-007-806

TTU-H2015-053-007-897

FEED SACK PRINTS – STRIPES AND PLAIDS

TTU-H2015-053-007-972   TTU-H2015-053-007-975   TTU-H2015-053-007-991   TTU-H2015-053-007-993

~NOTES~

CHAPTER ONE

1. Ruth Rhoades, "Feed Sacks in Georgia: Their Manufacture, Marketing, and Consumer Use," *Uncoverings*, no. 18, (1997): 130–31.

2. Rhoades, 125.

3. Loris Connolly, "Recycling Feed Sacks and Flour Bags: Thrifty Housewives or Marketing Success Story?" *Dress*, no. 19 (1992): 17.

4. Rhoades, 128.

5. Ibid.

6. Diana Twede, "The Cask Age: the Technology and History of Wooden Barrels," *Packaging Technology and Science*, no. 18 (2005): 253–64.

7. "What is a Hogshead? Barrels and Measurement in Colonial America," www.owlcation.com, accessed Nov. 27, 2017.

8. Rhoades, 123.

9. Twede, 253–64.

10. Kendra Brandes, "Feed Sack Fashion in Rural America: A Reflection of Culture," *The Online Journal of Rural Research and Policy,* no. 4 (1) (2009): 3.

11. Rhoades, 123.

12. "Cotton Sacks for Flour Win," *Dallas Morning News* (Dallas, Texas), July 27, 1932, p 3.

13. Rhoades, 123.

14. Ibid.

15. "History of Paper Packaging & Paper Carrier Bags," www.keenpa.com, accessed Nov. 27, 2017.

16. Rhoades, 124.

17. Cook, 13.

18. Ibid.

19. Beth Thorne Newcome and Joanna S. Nesselroad, Appendix D "Fairmont State College Textile Bag Collection," in Fawn Valentine, *West Virginia Quilts and Quiltmakers: Echoes From the Hills* (Athens, Ohio: Ohio University Press, 2000), 260.

20. Newcome and Nesselroad, 261.

21. John Hoye, *Staple Cotton Fabrics: Names, Descriptions, Finishes and Uses of Unbleached, Converted and Mill Finished Fabrics* (New York: McGraw Hill, 1942), 25.

22. Connolly, 22.

23. Rhoades, 130.

24. "Nicholson Co. to Use Cotton Bags to Help Present Cotton Slump," *Dallas Morning News* (Dallas, Texas), October 24, 1926, p 1.

25. "Why Cotton Isn't Used to Wrap Cotton," *Dallas Morning News* (Dallas, Texas), December 7, 1916, 14.

26. "New Uses for Cotton Sought," *Dallas Morning News* (Dallas, Texas), October 17, 1926, 1.

27. "Jute Cuts Out 750,000 Bales," *Dallas Morning News* (Dallas, Texas), September 4, 1927, 9.

28. "Food Bags Offer New Cotton Use," *Dallas Morning News* (Dallas, Texas), September 17, 1930, 25.

29. "Will Use Cotton Bags," *Dallas Morning News* (Dallas, Texas), December 6, 1926, 10.

30. "McDonald Urges More Cotton Use," *Dallas Morning News* (Dallas, Texas), April 10, 1931, 23.

31. "Cotton Bags used for Packing Nationally Known Product," *Dallas Morning News* (Dallas, Texas), September 27, 1931, 6.

32. "Wyatt's Orders Sugar Forwarded to Stores Packed in Cotton Bags," *Dallas Morning News* (Dallas, Texas), November 22, 1935, 2.

33. "Cotton Bag to Package Sugar Made," *Dallas Morning News* (Dallas, Texas), August 6, 1936, 7.

34. "New Uses of Cotton," *Dallas Morning News* (Dallas, Texas), August 11, 1936, 2.

35. "Cotton Bags Help to Sell Potatoes," *Dallas Morning News* (Dallas, Texas), September 17, 1932, 3.

36. "Cotton Bags Used for 500 Products," *Dallas Morning News* (Dallas, Texas), July 20, 1933, 3.

37. "Marc Anthony Sets Up Own Cotton Firm," *Dallas Morning News* (Dallas, Texas), June 4–6, 1939, 6.

38. "More Cotton Bags Urged by Anthony," *Dallas Morning News* (Dallas, Texas), June 19, 1939, 6.

39. "Cotton Displacing Bags for Wool," *Dallas Morning News* (Dallas, Texas), September 9, 1938, 17.

40. "Wants Cotton Bags Instead of Paper," *Dallas Morning News* (Dallas, Texas), June 16, 1938, 10.

41. "Use Cotton Bags in Shipping Fertilizer," *Dallas Morning News* (Dallas, Texas), February 20, 1927, 1.

42. "Cotton Sacks for Onions to Help Growers," *Dallas Morning News* (Dallas, Texas), May 6, 1936, 1.

43. "Line Industry Needs Onion Growers' Help," *Dallas Morning News* (Dallas, Texas), March 10, 1941, 10.

44. Gloria Nixon, *Rag Darlings; Dolls From the Feedsack Era* (Kansas City, Missouri: Kansas City Star Books, 2015), 11.

45. Ibid, 16.

46. Ibid, 7.

47. Ibid, 25.

48. Ibid, 31

49. Ibid.

50. Ibid, 33.

51. Ibid, 51–52.

52. "Chic Dresses Costing 3 Cents Displayed by Girls at Atlanta Cotton Style Show," *New York Times* (New York City, NY), August 6, 1933, N1.

53. "Wallace Sees Feed Bag Dresses," *Dallas Morning News* (Dallas, Texas), August 10, 1933, 8.

54. Rhoades, 125.

55. Connolly, 21.

56. Ibid, 22.

57. "Why Not This Paisley Pattern? Biddy Won't Know Difference," *Dallas Morning News* (Dallas, Texas), July 13, 1944, 1.

58. "Flour Sack Fashions," *Dallas Morning News* (Dallas, Texas), July 28, 1949, 13.

59. "Bag Clothing Again; Wants to Be Certain," *Dallas Morning News* (Dallas, Texas), February 28, 1952, 3.

CHAPTER TWO

1. Loris Connolly, "Recycling Feed Sacks and Flour Bags: Thrifty Housewives or Marketing Success Story?" *Dress*, no. 19 (1992): 17.

2. Anna Lue Cook, *Textile Bags Identification and Value Guide (The Feeding and Clothing of America)* (Florence, Alabama: Books Americana, 1990), 94.

3. "'Farm-Tested' Removed from Larro Label," *The Modern Millwheel*, February, 1944, 2.

4. "'Farm-tested' Trademark Returns!" *Larro Feeder*, 1945, 1.

5. "'Farm-tested' Label on Feed Bags Again," *The Modern Millwheel*, March, 1947, 1.

6. Linzee Kull McCray, *Feed Sacks: The Colorful History of a Frugal Fabric*, (Calgary, Alberta, Canada: Uppercase Publishing Inc, 2016), 213–14.

CHAPTER THREE

1. *Sewing with Flour Bags*, circa 1926–27, 16–17.

2. Virginia Gunn, "McCall's Role in the Early Twentieth-Century Quilt Revival," *Uncoverings* 31, (2010), 17–18.

3. Rose Marie Werner, e-mail correspondence with the author, October 30, 2018.

4. Rose Marie Werner, e-mail correspondence with the author, October 30, 2018.

5. Vintage Household Chores Set, http://tipnut.com/household-chores-towels/ from www.tipnut.com, accessed September, 2016.

6. Rose Marie Werner, e-mail correspondence with author, October 30, 2018.

7. Rose Marie Werner, e-mail correspondence with the author, October 30, 2018.

CHAPTER FOUR

1. Edward Eyre Hunt, *War Bread: A Personal Narrative of the War and Relief in Belgium* (New York: Henry Holt and Company, 1916), 229.

2. Ibid, 229–30.

3. Charlotte Kellogg, *Women of Belgium; Turning Tragedy to Triumph* (New York and London: Funk & Wagnalls Company, 1917), 154.

4. Hunt, 296.

5. "The Apotheosis of the Flour Bag: How a Belgian Expresses Gratitude," *The Modern Priscilla*, February 1917, 5.

6. World War I & the Rockefeller Foundation website https://rockfound.rockarch.org/world-war-i-the-rf accesses July 2, 2018.

7. Hunt, 297.

8. Ibid, 301.

9. Ibid, 165.

10. See Annelien van Kempen's website http://www.annelienvankempen.nl/Resources/Een%20Canadese%20meelzak%20in%20TRC.pdf

CHAPTER FIVE

1. Marjory L. Walker, National Cotton Council Staff, email message to author, July 25, 2017 referencing information published in the Council's *The First Forty Years, A History of the National Cotton Council of America, 1939–1979.*

2. An earlier version of this research was published in the Winter 2017–18, Issue 132 of *Blanket Statements*, a News Publication of the American Quilt Study Group.

3. Loris Connolly, "Recycling Feed Sacks and Flour Bags: Thrifty Housewives or Marketing Success Story?" *Dress*, no. 19 (1992): 18.

4. Shirley Eagan, (1990) "Women's Work, Never Done; West Virginia Farm Women, 1880s–1920s," *West Virginia History*, published by West Virginia Archives and History, 49, (1990): 21–36.

5. The Household Science Institute. *Sewing with Flour Bags.* Circa 1926.

6. Marjory L. Walker, National Cotton Council Staff, email message to author, July 25, 2017 referencing information published in the Council's *The First Forty Years, A History of the National Cotton Council of America, 1939–1979.*

7. Ibid.

8. Pat L. Nickols, "The Use of Cotton Sacks in Quiltmaking, *Uncoverings*, no 9,m (1988): 65–6.

9. See the back of TTU-H1979-180, which is pictured in the chapter on quilts.

10. The Household Science Institute, *Sewing with Flour Bags.* Circa 1926, 2.

11. The Textile Bag Manufacturers Association. *Sewing with Cotton Bags*, 6-1937, 3.

12. National Cotton Council, *Bag Magic for Home Sewing*, c 1944, 2.

13. National Cotton Council, *Smart Sewing with Cotton Bags*, c 1947, 3.

14. National Cotton Council, *Sew Easy with Cotton Bags*, 1948, 3.

15. National Cotton Council, *1952 Pattern Service for Sewing with Cotton Bags,* 1952, 2.

16. National Cotton Council, *1953 Pattern Service for Sewing with Cotton Bags*, 1953, 4.

17. National Cotton Council, *1954 Idea Book for Sewing with Cotton Bags*, 1954, 4.

18. National Cotton Council, *Ideas for Sewing with Cotton Bags*, c 1955, 4.

19. "Pillowcases New Sensation of Cotton Bag Business," *Dallas Morning News* (Dallas, Texas), October 4, 1953, 9.

20. Robert McG Thomas, Jr., "James Shapiro, 85, Innovator in the Home Sewing Industry," *The New York Times* (New York City, New York), June 3, 1995.

21. "Mary Lynch Lincoln," Lancasteronline.com, July 8, 2008.

22. Edmond Lincoln interview with author, August 9, 2017.

23. *Ideas for Sewing with Cotton Bags,* centerfold, 1955.

24. E-mail from National Cotton Council Staff, Marjory L. Walker to author, July 25, 2017 referencing information published in the Council's *The First Forty Years, A History of the National Cotton Council of America, 1939–1979*.

25. Connolly, 22.

26. Beth Thorne Newcome and Joanna S. Nesselroad, Appendix D "Fairmont State College Textile Bag Collection," in Fawn Valentine, *West Virginia Quilts and Quiltmakers: Echoes From the Hills* (Athens, Ohio: Ohio University Press, 2000), 261–2.

27. "New Use for Sugar Sacks," *New York Times* (New York City, New York), December 6, 1944, 20.

28. "A Bag of Tricks for Home Sewing," Consumer Education Division Sears, Roebuck and Company in Co-operation with the National Cotton Council, c 1942–43, p 25.

29. https://www.vintagestitching.com/pages/dating-vintage-patterns, accessed July 7, 2018.

30. http://www.cemetarian.com/index.php?pr=Simplicity, accessed July 7, 2018.

31. "Cotton: A Double Life," *Time* Magazine, January 31, 1949, 58 & 60.

32. Ibid, p. 60

33. Advertisement, *Feedstuffs* 19 (June 21, 1947); 17.

34. Wright, Sarah Bliss, "Alabama Cotton and Bemis Bags," *Uncoverings* 2013, 34: 148.

35. "Art Show Underlines New Styles," *Dallas Morning News* (Dallas, Texas), April 26, 1958, 6.

CHAPTER SIX

1. Minnie Church, "What to do with the Sacks," *American Cookery*, May 1921, 207.

2. "Women: Foul Rumor," *Time* Magazine, March 11, 1946.

3. "Recipes Go Abroad with Farm Women," *New York Times* (New York City, New York), August 24, 1950, 29.

4. Jennifer Lynn Banning, "Feed Sack Fashions in South Louisiana, 1949–1968; The Use of Commodity Bags in Garment Construction," (doctoral dissertation, Louisiana State University, 2005), ix.

5. Fisher, Linda, e-mail correspondence with author, November 1, 2018.

6. Adrosko, Rita, "The Fashion's in the Bag: Recycling Feed, Flour, and Sugar Sacks During the Middle Decades of the 20th Century," *Textile Society of America Symposium Proceedings* 1992): 130.

CHAPTER SEVEN

1. Steve Harmon, "Throwback Thursday: East Sherman, 1891," *The Hutchinson News* (Hutchinson, Kansas) July 20, 2017.

2. "The Biggest Little Board of Trade," *The Southwestern Grain and Flour Journal* 7, no. 12 (January, 1914): 12.

CHAPTER EIGHT

1. Loris Connolly, "Recycling Feed Sacks and Flour Bags: Thrifty Housewives or Marketing Success Story?" *Dress*, no. 19 (1992): 21.

2. Ibid.

3. Jennifer Lynn Banning, "Feed Sack Fashions in South Louisiana, 1949–1968; The Use of Commodity Bags in Garment Construction," (doctoral dissertation, Louisiana State University, 2005), p 15.

4. Connolly, 22.

5. Ruth Rhoades, "Feed Sacks in Georgia: Their Manufacture, Marketing, and Consumer Use," *Uncoverings*, no. 18, (1997): 130–31.

6. Banning, 16. See also Eileen Jahnke Trestain, *Dating fabrics: A color guide 1800–1960* (Paducah, KY: American Quilter's Society, 1998), 165; and Beth Thorne Newcome and Joanna S. Nesselroad, Appendix D "Fairmont State College Textile Bag Collection," in Fawn Valentine, *West Virginia Quilts and Quiltmakers: Echoes From the Hills* (Athens, Ohio: Ohio University Press, 2000) 260–67.

7. Anna Lue Cook, *Textile Bags Identification and Value Guide (The Feeding and Clothing of America)* (Florence, Alabama: Books Americana, 1990), 14.

8. Cook, 13.

9. Newcome and Nesselroad, 261.

10. Ibid, 262.

11. Ibid.

12. Ibid.

13. A lightweight, sheer cotton fabric having at least two warp threads thrown into relief to form fine cords.

14. Made of a weaving style that is characterized by raised parallel cords or geometric designs in the fabric.

15. A dense plain woven fabric woven historically of wool but in the mid twentieth century also of cotton.

16. A thin puckered all cotton fabric where the puckering forms stripes or checks.

17. "*A Bag of Tricks for Home Sewing*," Consumer Education Division, Sears, Roebuck and Company in cooperation with the National Cotton Council of America, c 1942–43, p 27.

18. Newcome and Nesselroad, 261.

19. Ibid.

20. *Needle Magic with Cotton Bags* (Memphis, Tennessee: National Cotton Council, c 1949–50), 23.

21. Newcome and Nesselroad, 261.

22. Notes from Charlotte Williams with her 2015 donation to the Museum of Texas Tech University, Clothing and Textiles Division Accession File TTU-H2015-013, March, 2015.

23. LuAnn Jones, *Mama Learned Us to Work: Farm Women in the New South* (Chapel Hill and London: University of North Carolina Press, 2002), 178.

24. Jones, 179.

25. Newcome and Nesselroad, 263.

26. "Textile Plants Face Vast War Change-Over," *Dallas Morning News* (Dallas, Texas), April 22, 1942, 7.

27. "War Compels Conservation of Feed Sacks," *Dallas Morning News* (Dallas, Texas), April 20, 1942, 4.

28. "Women, Girls of Farm Clubs Going to Camp," *Dallas Morning News* (Dallas, Texas), July 18, 1942, 3.

29. "Feed-Sack Styles," *Dallas Morning News* (Dallas, Texas), August 12, 1944, 9. And "Cotton, Synthetics Topic for Council," *Dallas Morning News* (Dallas Texas), September 17, 1944, 4.

30. Margaret Powel, "From Feed Sack to Clothes Rack: The Use of Commodity Textile Bags in American Households from 1890–1960," *Textile Society of America Symposium Proceedings* 9-2012: 9.

31. "High School Girl Wardrobe Designed for Less than $50," *Dallas Morning News* (Dallas, Texas), August 16, 1942, 8.

32. "Rummage Sale Ain't Nothin' when These Folks Scramble to Plank Down Cash for Baby Chicks," *Dallas Morning News* (Dallas, Texas), April 18, 1943, 1.

33. "HDC (Home Demonstration Club) Week Observed with Gifts of Clothing," *Dallas Morning News* (Dallas, Texas), May 5, 1946, 1.

34. "Flour Sack Not Sad," *Dallas Morning News* (Dallas, Texas), July 16, 1947, 1.

35. "Delegates to Regional 4-H Point to Better Farm Living, *Atlanta Daily World* (Atlanta, Georgia), July 12, 1949, 4.

36. "Pillowcases New Sensation of Cotton Bag Business," *Dallas Morning News* (Dallas, Texas), October 4, 1953, 9.

37. The Dallas plant built in 1906 was located at 4301 South Fitzhugh.

38. "Pillowcases New Sensation of Cotton Bag Business," *Dallas Morning News* (Dallas, Texas), October 4, 1953, 9.

39. Cook, 13.

40. Wright, Sarah Bliss, "Alabama Cotton and Bemis Bags," *Uncoverings* 2013 34: 151–2.

41. "Burlap Sandbags to be Replaced by Synthetics," *Dallas Morning News* (Dallas, Texas), August 18, 1966, 23.

❧ BIBLIOGRAPHY ❧

BOOKS

Brackman, Barbara. *Encyclopedia of Pieced Quilt Block Patterns*. Lawrence, Kansas: Prairie Flower Publishing, 1984.

Cook, Anna Lue. *Identification and Value Guide to Textile Bags (The Feeding and Clothing of America)*. Florence, Alabama: Books Americana, 1990.

Hoye, John. *Staple Cotton Fabrics: Names, Descriptions, Finishes and Uses of Unbleached, Converted and Mill Finished Fabrics*. New York: McGraw Hill, 1942.

Hunt, Edward Eyre. *War Bread: A Personal Narrative of the War and Relief in Belgium*. New York: Henry Holt and Company, 1916.

Jones, Lu Ann. *Mama Learned Us to Work: Farm Women in the New South*. Chapel Hill & London: The University of North Carolina Press, 2002.

Kellogg, Charlotte. *Women of Belgium; Turning Tragedy to Triumph*. New York and London: Funk & Wagnalls Company, 1917.

McCray, Linzee Kull. *Feed Sacks: The Colourful History of a Frugal Fabric*. Calgary, Alberta, Canada: Uppercase Publishing Inc., 2016.

Nixon, Gloria. *Rag Darlings*. Kansas City, Missouri: Kansas City Star Books, 2015.

Trestain, Eileen Jahnke. *Dating Fabrics: A Color Guide 1800–1960*. Paducah, KY: American Quilter's Society, 1998.

Valentine, Fawn. *West Virginia Quilts and Quiltmakers: Echoes From the Hills*. Athens, Ohio: Ohio University Press, 2000.

Waldvogel, Merikay. *Soft Covers for Hard Times: Quiltmaking and the Great Depression*. Nashville, TN: Rutledge Hill Press, 1990.

DISSERTATIONS

Banning, Jennifer Lynn. "Feed Sack Fashions in South Louisiana, 1949–1968; The Use of Commodity Bags in Garment Construction," Doctoral dissertation, Louisiana State University, 2005.

JOURNALS

Adrosko, Rita. "The Fashion's in the Bag: Recycling Feed, Flour, and Sugar Sacks During the Middle Decades of the 20th Century." *Textile Society of America Symposium Proceedings*, (1992).

"The Biggest Little Board of Trade." *The Southwestern Grain and Flour Journal* 7, no. 12 (January, 1914): 12.

Brandes, Kendra. "Feed Sack Fashion in Rural America: A Reflection of Culture." *Online Journal of Rural Research & Policy* 4, no. 1 (2009).

Church, Minnie. "What to do with the Sacks." *American Cookery*, May 1921.

Connolly, Loris. "Recycling Feed Sacks and Flour Bags: Thrifty Housewives or Marketing Success Story? *Dress* 19, (1992).

Eagan, Shirley C. (1990). "Women's Work, Never Done; West Virginia Farm Women, 1880s–1920s." *West Virginia History,* published by West Virginia Archives and History, 49, (1990).

Gunn, Virginia. "McCall's Role in the Early Twentieth-Century Quilt Revival." *Uncoverings* 31, (2010).

Nickols, Pat L. "The Use of Cotton Sacks in Quiltmaking." *Uncoverings* 9, (1988).

Powell, Margaret. "From Feed Sack to Clothes Rack: The Use of Commodity Textile Bags in American Households from 18901960." *Textile Society of America Symposium Proceedings* (9-2012).

Rhoades, Ruth. "Feed Sacks in Georgia: Their Manufacture, Marketing and Consumer Use." *Uncoverings* 18, (1997).

Twede, Diana. "The Cask Age: the Technology and History of Wooden Barrels." *Packaging Technology and Science* 18, (2015).

Wright, Sarah Bliss. "Alabama Cotton and Bemis Bags." *Uncoverings* 24, (2013).

NEWSPAPERS

"Art Show Underlines New Styles," *Dallas Morning News* (Dallas, Texas), April 26, 1958.

"Bag Clothing Again; Wants to Be Certain," *Dallas Morning News* (Dallas, Texas), February 28, 1952.

"Burlap Sandbags to be Replaced by Synthetics," *Dallas Morning News* (Dallas, Texas), August 18, 1966.

"Chic Dresses Costing 3 Cents Displayed by Girls at Atlanta Cotton Style Show," *New York Times* (New York City, NY), August 6, 1933.

"Cotton: A Double Life," *Time* Magazine, January 31, 1949, 58 & 60.

"Cotton Bag to Package Sugar Made," *Dallas Morning News* (Dallas, Texas), August 6, 1936.

"Cotton Bags Help to Sell Potatoes," *Dallas Morning News* (Dallas, Texas), September 17, 1932.

"Cotton Bags Used for 500 Products," *Dallas Morning News* (Dallas, Texas), July 20, 1933.

"Cotton Bags used for Packing Nationally Known Product," *Dallas Morning News* (Dallas, Texas), September 27, 1931.

"Cotton Displacing Bags for Wool," *Dallas Morning News* (Dallas, Texas), September 9, 1938.

"Cotton Sacks for Flour Win," *Dallas Morning News* (Dallas, Texas), July 27, 1932.

"Cotton Sacks for Onions to Help Growers," *Dallas Morning News* (Dallas, Texas), May 6, 1936.

"Cotton, Synthetics Topic for Council," *Dallas Morning News* (Dallas Texas), September 17, 1944.

"Delegates to Regional 4-H Point to Better Farm Living, *Atlanta Daily World* (Atlanta, Georgia), July 12, 1949.

"Feed-Sack Styles," *Dallas Morning News* (Dallas, Texas), August 12, 1944.

"Flour Sack Fashions," *Dallas Morning News* (Dallas, Texas), July 28, 1949.

"Flour Sack Not Sad," *Dallas Morning News* (Dallas, Texas), July 16, 1947.

"Food Bags Offer New Cotton Use," *Dallas Morning News* (Dallas, Texas), September 17, 1930.

Harmon, Steve. "Throwback Thursday: East Sherman, 1891," *The Hutchinson News* (Hutchinson, Kansas) July 20, 2017.

"HDC (Home Demonstration Club) Week Observed with Gifts of Clothing," *Dallas Morning News* (Dallas, Texas), May 5, 1946.

"High School Girl Wardrobe Designed for Less than $50," *Dallas Morning News* (Dallas, Texas), August 16, 1942.

"Jute Cuts Out 750,000 Bales," *Dallas Morning News* (Dallas, Texas), September 4, 1927.

"Kerosene Removes Printing," *Dallas Morning News* (Dallas, Texas), November 1, 1963.

"Line Industry Needs Onion Growers' Help," *Dallas Morning News* (Dallas, Texas), March 10, 1941.

"Marc Anthony Sets Up Own Cotton Firm," *Dallas Morning News* (Dallas, Texas), June 4 6, 1939.

"McDonald Urges More Cotton Use," *Dallas Morning News* (Dallas, Texas), April 10, 1931.

"More Cotton Bags Urged By Anthony," *Dallas Morning News* (Dallas, Texas), June 19, 1939.

"Nicholson Co. to Use Cotton Bags to Help Present Cotton Slump," *Dallas Morning News* (Dallas, Texas), October 24, 1926.

"New Use for Sugar Sacks," *New York Times* (New York City, New York), December 6, 1944.

"New Uses for Cotton Sought," *Dallas Morning News* (Dallas, Texas), October 17, 1926.

"New Uses of Cotton," *Dallas Morning News* (Dallas, Texas), August 11, 1936.

"Pillowcases New Sensation of Cotton Bag Business," *Dallas Morning News* (Dallas, Texas), October 4, 1953.

"Recipes Go Abroad with Farm Women," New York Times (New York City, New York), August 24, 1950.

"Rummage Sale Ain't Nothin' when These Folks Scramble to Plank Down Cash for Baby Chicks," *Dallas Morning News* (Dallas, Texas), April 18, 1943.

"Textile Plants Face Vast War Change-Over," *Dallas Morning News* (Dallas, Texas), April 22, 1942.

"The Apotheosis of the Flour Bag: How a Belgian Expresses Gratitude," *The Modern Priscilla*, February 1917.

Thomas, Jr., Robert McG.. "James Shapiro, 85, Innovator in the Home Sewing Industry," *The New York Times* (New York City, New York), June 3, 1995.

"Uncle Sam Chided for Heavy Bite," *Dallas Morning News* (Dallas, Texas), April 14, 1957.

"Use Cotton Bags in Shipping Fertilizer," *Dallas Morning News* (Dallas, Texas), February 20, 1927.

"Wallace Sees Feed Bag Dresses," *Dallas Morning News* (Dallas, Texas), August 10, 1933.

"Wants Cotton Bags Instead of Paper," *Dallas Morning News* (Dallas, Texas), June 16, 1938.

"War Compels Conservation of Feed Sacks," Dallas Morning News (Dallas Texas), April 20, 1942.

"Why Cotton Isn't Used to Wrap Cotton," *Dallas Morning News* (Dallas, Texas), December 7, 1916.

"Why Not This Paisley Pattern? Biddy Won't Know Difference," *Dallas Morning News* (Dallas, Texas), July 13, 1944.

"Will Use Cotton Bags," *Dallas Morning News* (Dallas, Texas), December 6, 1926.

"Women: Foul Rumor," *Time Magazine*, March 11, 1946.

"Women, Girls of Farm Clubs Going to Camp," *Dallas Morning News* (Dallas, Texas), July 18, 1942.

"Wyatt's Orders Sugar Forwarded to Stores Packed in Cotton Bags," *Dallas Morning News* (Dallas, Texas), November 22, 1935.

WEBSITES

Annelien van Kempen's website http://www.anne lienvankempen.nl/Resources/Een%20Canadese%20 meelzak%20in%20TRC.pdf

"History of Paper Packaging & Paper Carrier Bags," www.keenpa.com, accessed Nov. 27, 2017.

"Vintage Household Chores Set," http://tipnut.com /household-chores-towels/ from www.tipnut.com, accessed September, 2016.

"What is a Hogshead? Barrels and Measurement in Colonial America," www.owlcation.com, accessed Nov. 27, 2017.

World War I & the Rockefeller Foundation website https://rockfound.rockarch.org/world-war-i-the-rf accessed July 2, 2018.

✦INDEX✦

⤳ABOUT THE AUTHOR⤳

Curator of Clothing and Textiles at the Museum of Texas Tech University since 2014, Dr. Marian Ann J. Montgomery cares for over 33,000 objects, likely the largest collection of this type of material at a University in the United States. She is a quilt historian having published through the American Quilt Study Group. Sought out as an entertaining and informative speaker, Dr. Montgomery enjoys making history come alive through the objects. She received the 2019 Bybee Scholar award for her work in promoting and preserving the art of quilting. Dr. Montgomery earned her Ph.D. in fashion and textile history/museum administration from New York University through studies in the Costume Institute and Textile Study Room at the Metropolitan Museum of Art. She resides in Lubbock with her husband and dog where she quilts in her spare time.

The publication of this catalogue is generously supported by United Notions / Moda Fabrics and The CH Foundation.

CPSIA information can be obtained
at www.ICGtesting.com
Printed in the USA
LVHW071702020819
626320LV00013B/227/P